Radioactive Substances

Marie Curie

DOVER PUBLICATIONS, INC.
Mineola, New York

Published in the United Kingdom by David & Charles, Brunel House, Forde Close, Newton Abbot, Devon TQ12 4PU.

Bibliographical Note

Radioactive Substances was Mme. Curie's doctoral dissertation, presented at the Sorbonne in 1903. It was first published as "Recherches sur les Substances Radioactives" in the journal *Annales de chimie et de physiques*, and as a book under the same title by Gauthier-Villars, Paris, in 1903. A second, revised edition was published by Gauthier-Villars, Paris, in 1904. An English translation had appeared under the title "Radio-Active Substances" in the journal *Chemical News* Office, London, and by Van Nostrand, New York, in 1904.

This Dover edition, first published in 2002, is an unabridged republication of the 1904 edition published as *Radioactive Substances* by Philosophical Library, New York, in 1961. The introduction and frontispiece from that edition have been omitted.

Library of Congress Cataloging-in-Publication Data

Curie, Marie, 1867–1934.
 [Recherches sur les Substances Radioactives. English]
 Radioactive Substances / Curie, Marie.–Dover ed.
     p. cm.
     Previously published: Radio-Active Substances. 2nd ed. London: Chemical News Office,1904.
     Includes bibliographical references and index.
     ISBN 0-486-42550-9 (pbk.)
     1. Radioactivity. 2. Radioactive Substances. I. Title.

QC795 .C82313 2002b
539.7'52—dc21

2002025614

Manufactured in the United States of America
Dover Publications, Inc., 31 East 2nd Street, Mineola, N.Y. 11501

Radioactive Substances

RADIO-ACTIVE SUBSTANCES.

INTRODUCTION.

THE object of the present work is the publication of researches which I have been carrying on for more than four years on radio-active bodies. I began these researches by a study of the phosphorescence of uranium, discovered by M. Becquerel. The results to which I was led by this work promised to afford so interesting a field that M. Curie put aside the work on which he was engaged, and joined me, our object being the extraction of new radio-active substances and the further study of their properties.

Since the commencement of our research we thought it well to hand over specimens of the substances, discovered and prepared by ourselves, to certain physicists, in the first place to M. Becquerel, to whom is due the discovery of the uranium rays. In this way we ourselves facilitated the research by others besides ourselves on the new radio-active bodies. At the termination of our first publications, M. Giesel, in Germany, also began to prepare these substances, and passed on specimens of them to several German scientists. Finally, these substances were placed on sale in France and Germany, and the subject growing in importance gave rise to a scientific movement, such that numerous memoirs have appeared, and are constantly appearing on radio-active bodies, principally abroad. The results of the various French and foreign researches are necessarily confused, as is the case with all new subjects in course of investigation, the aspect of the question becoming modified from day to day.

From the chemical point of view, however, one point is definitely established :—*i.e.*, the existence of a new element, strongly radio-active, viz., radium. The preparation of the pure chloride of radium and the determination of the atomic weight of radium form the chief part of my own work. Whilst this work adds to the elements actually known with certainty a new element with very curious properties, a new method of chemical research is at the same time established and justified. This method, based on the consideration of radio-activity as an atomic property of matter, is just that which enabled M. Curie and myself to discover the existence of radium.

If, from the chemical point of view, the question that we undertook primarily may be looked upon as solved, the study of the physical properties of the radio-active bodies is in full evolution. Certain important points have been established, but a large number of the conclusions are still of a provisional character. This is not surprising when we consider the complexity of the phenomena due to radio-activity, and the differences existing between the various radio-active substances. The researches of physicists on these substances constantly meet and overlap. Whilst endeavouring to keep strictly to the limits of this work and to publish my individual research only, I have been obliged at the same time to mention results of other researches, the knowledge of which is indispensable.

I desired, moreover, to make this work an inclusive survey of the actual position of the question.

I indicate at the end the particular questions with which I am specially concerned, and those which I investigated in conjunction with M. Curie.

I carried on the work in the laboratories of the School of Physics and Chemistry in Paris, with the permission of Schützenberger, late Director of the School, and M. Lauth, actual Director. I take this opportunity of expressing my gratitude for the kind hospitality received in this school.

HISTORICAL.

The discovery of the phenomena of radio-activity is connected with researches followed, since the discovery of the Röntgen rays, upon the photographic effects of phosphorescent and fluorescent substances.

The first tubes for producing Röntgen rays were without the metallic anti-cathode. The source of the Röntgen rays was the glass surface impinged upon by the cathode rays; this surface was at the same time actively fluorescent. The question then was whether the emission of Röntgen rays necessarily accompanied the production of fluorescence, whatever might be the cause of the latter. This idea was first enunciated by M. Henri Poincaré.

Shortly afterwards, M. Henry announced that he had obtained photographic impressions through black paper by means of phosphorescent zinc sulphide. M. Niewenglowski obtained the same phenomenon with calcium sulphide exposed to the light. Finally, M. Troost obtained strong photographic impressions with zinc sulphide artificially phosphorescent acting across black paper and thick cardboard.

The experiences just cited have not been reproduced, in spite of numerous attempts to this end. It cannot therefore be considered as proved that zinc sulphide and calcium sulphide are capable of emitting, under the action of light, invisible rays which traverse black paper and act on photographic plates.

M. Becquerel has made similar experiments on the salts of uranium, some of which are fluorescent.

He obtained photographic impressions through black paper with the double sulphate of uranium and potassium.

M. Becquerel at first believed that this salt, which is fluorescent, behaved like the sulphides of zinc and calcium in the experiments of MM. Henry, Niewenglowski, and Troost. But the conclusion of his experiments showed that the phenomenon observed was in no way related to the fluorescence. It is not necessary that the salt should be fluorescent; further, uranium and all its compounds, fluorescent or not, act in the same manner, and metallic uranium is the most active. M. Becquerel finally found that by placing uranium compounds in complete darkness, they continue acting on photographic plates through black paper for years. M. Becquerel allows that uranium and its compounds emit peculiar rays—uranium rays. He proved that these rays can penetrate thin metallic screens, and that they discharge electrified bodies. He also made experiments from which he concluded that uranium rays undergo reflection, refraction, and polarisation.

The work of other physicists (Elster and Geitel, Lord Kelvin, Schmidt, Rutherford, Beattie, and Smoluchowski) confirms and extends the results of the researches of M. Becquerel, with the exception of those relating to the reflection, refraction, and polarisation of uranium rays, which in this respect behave like Röntgen rays, as has been recognised first by Mr. Rutherford and then by M. Becquerel himself.

CHAPTER I.

RADIO-ACTIVITY OF URANIUM AND THORIUM. RADIO-ACTIVE MINERALS.

Becquerel Rays.—The uranium rays discovered by M. Becquerel act upon photographic plates screened from the light; they can penetrate all solid, liquid, and gaseous substances, provided that the thickness is sufficiently reduced in passing through a gas, they cause it to become a feeble conductor of electricity.

These properties of the uranium compounds are not due to any known cause. The radiation seems to be spontaneous; it loses nothing in intensity, even on keeping the compounds in complete darkness for several years; hence there is no question of the phosphorescence being specially produced by light.

The spontaneity and persistence of the uranium radiation appear as a quite unique physical phenomenon. M. Becquerel kept a piece of uranium for several years in the dark, and he has affirmed that at the end of this time the action upon a photographic plate had not sensibly altered. MM. Elster and Geitel made a similar experiment, and also found the action to remain constant.

I measured the intensity of radiation of uranium by the effect of this radiation on the conductivity of air. The method of measurement will be explained later. I also obtained figures which prove the persistence of radiation within the limits of accuracy of the experiments.

For these measurements a metallic plate was used covered with a layer of powdered uranium; this plate was not otherwise kept in the dark; this precaution, according to the experimenters already quoted, being of no importance. The number of measurements taken with this plate is very great, and they actually extend over a period of five years.

Some researches were conducted to discover whether other substances were capable of acting similarly to the uranium compounds. M. Schmidt was the first to publish that thorium and its compounds possess exactly the same property. A similar research, made contemporaneously, gave me the same result. I published this not knowing at the time of Schmidt's publication.

We shall say that uranium, thorium, and their compounds emit *Becquerel rays.* I have called *radio-active* those substances which generate emissions of this nature. This name has since been adopted generally.

In their photographic and electric effects, the Becquerel rays approximate to the Röntgen rays. They also, like the latter, possess the faculty of penetrating all matter. But their capacity for penetration is very different; the rays of uranium and of thorium are arrested by some millimetres of solid matter, and cannot traverse in air a distance greater than a few centimetres; this at least is the case for the greater part of the radiation.

The researches of different physicists, and primarily of Mr. Rutherford, have shown that the Becquerel rays undergo neither regular reflection, nor refraction, nor polarisation.

The feeble penetrating power of uranium and thorium rays would point to their similarity to the secondary rays produced by the Röntgen rays, and which have been investigated by M. Sagnac, rather than to the Röntgen rays themselves.

For the rest, the Becquerel rays might be classified as cathode rays propagated in the air. It is now known that these different analogies are all legitimate.

Measurement of the Intensity of Radiation.

The method employed consists in measuring the conductivity acquired by air under the action of radio-active bodies; this method possesses the advantage of being rapid and of furnishing figures which are comparable. The apparatus employed by me for the purpose consists essentially of a plate condenser, A B (Fig. 1). The active body, finely powered, is spread over the plate B, making the air between the plates a conductor. In order to measure the conductivity, the plate B

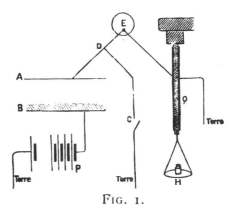

Fig. 1.

is raised to a high potential by connecting it with one pole of a battery of small accumulators, P, of which the other pole is connected to earth. The plate A being maintained at the potential of the earth by the connection C D, an electric current is set up between the two plates. The potential of plate A is recorded by an electrometer, E. If the earth connection be broken at C, the plate A becomes charged, and this charge causes a deflection of the electrometer. The velocity of the deflection is proportional to the intensity of the current, and serves to measure the latter.

But a preferable method of measurement is that of compensating the charge on plate A, so as to cause no deflection of the electrometer. The charges in question are extremely weak; they may be compensated by means of a quartz electric balance, Q, one sheath of which is connected to plate A and the other to earth. The quartz lamina is subjected to a known tension, produced by placing weights in a plate, π; the tension is produced progressively, and has the effect of generating progressively a known quantity of electricity during the time observed. The operation can be so regulated that, at each instant, there is compensation between the quantity of electricity that traverses the condenser and that of the opposite kind furnished by the quartz. In this way, the quantity of electricity passing through the condenser for a given time, *i.e.*, the *intensity of the current*, can be measured *in absolute units*. The measurement is independent of the sensitiveness of the electrometer.

In carrying out a certain number of measurements of this kind, it is seen that radio-activity is a phenomenon capable of being measured with a certain accuracy. It varies little with temperature; it is scarcely affected by variations in the temperature of the surroundings; it is not influenced by incandescence of the active substance. The intensity of the current which traverses the condenser increases with the surface of the plates. For a given condenser and a given substance the current increases with the difference of potential between the plates, with the pressure of the gas which fills the condenser, and with the distance of the plates (provided this distance be not too great in comparison with the diameter). In every case, for great differences of potential the current attains a limiting value, which is practically constant. This is the *current of saturation*, or *limiting current*. Similarly, for a certain sufficiently great distance between the plates the current hardly varies any longer with the distance. It is the current obtained under these conditions that was taken as the measure of radioactivity in my researches, the condenser being placed in air at atmospheric pressure.

I append curves which represent the intensity of the current as a function of the field established between the plates for two different plate distances. Plate B was covered with a thin layer of powdered metallic uranium; plate A, connected with the electrometer, was provided with a guard-ring.

Fig. 2 shows that the intensity of the current becomes constant for high potential differences between the plates. Fig. 3 represents the same curves on another scale, and

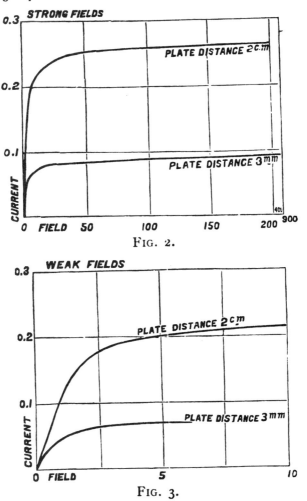

FIG. 2.

FIG. 3.

comprehends only relative results for small differences of potential. At the origin, the curve is rectilinear; the ratio of the intensity of the current to the difference of potential is constant for weak forces, and represents the initial conduction between the plates. Two important characteristic constants of the observed phenomenon are therefore to be recognised:—(1) The *initial conduction* for small differences of potential; (2) the *limiting current* for great potential differences. The limiting current has been adopted as the measure of the radio-activity.

Besides the difference of potential established between he two plates, there exists between them an electromotive force of contact, and these two sources of current combine their effects; for this reason, the absolute value of the intensity of the current changes with the sign of the external difference of potential. In every case, for considerable potential differences, the effect of the electromotive force of contact is negligible, and the intensity of the current is therefore the same whatever be the direction of the field between the plates.

The investigation of the conductivity of air and other gases subjected to the action of Becquerel rays has been undertaken by several physicists. A very complete research upon the subject has been published by Mr. Rutherford.

The laws of the conductivity produced in gases by the Becquerel rays are the same as those found for the Röntgen rays. The mechanics of the phenomenon appear to be the same in both cases. The theory of ionisation of the gases by the action of the Röntgen or Becquerel rays agrees well with the observed facts. This theory will not be put forward here. I will merely record the results to which they point:—

Firstly, the number of ions produced per second in the gas is considered proportional to the energy of radiation absorbed by the gas.

Secondly, in order to obtain the limiting current relatively to a given radiation, it is necessary, on the one hand, to cause complete absorption of this radiation by the gas by employing a sufficient mass of it; on the other hand, it is necessary for the production of the current to use all the ions generated by establishing an electric field of such strength that the number of the ions which recombine may be a negligible fraction of the total number of ions produced in the same time, most of which are carried by the current to the electrodes. The strength of the electric field necessary to give this result is proportional to the amount of ionisation.

According to the recent researches of Mr. Townsend, the

phenomenon is more complex when the pressure of the gas is low. At first the current appears to approach to a constant limiting value with increasing difference of potential; but after a certain point has been reached, the current begins again to increase with the field, and with very great rapidity. Mr. Townsend ascribes this increase to a new ionisation produced by the ions themselves when, under the action of the electric field, they acquire a velocity such that a molecule of gas encountering one of them becomes broken down into its constituent ions. A strong electric field and a low pressure are favourable to the production of this ionisation by ions already present, and, as soon as the action is set up, the intensity of the current increases uniformly with the field between the plates. The limiting current could, therefore, only be obtained under conditions of ionisation of which the intensity does not exceed a certain value, and in such a manner that saturation corresponds to fields in which, from multiplicity of ions, ionisation can no longer take place. This condition has occurred in my experiments.

The order of magnitude of the saturation currents obtained with uranium compounds is 10^{-11} ampères for a condenser in which the plates have a diameter of 8 c.m., and are at a distance of 3 c.m. Thorium compounds give rise to currents of the same order of magnitude, and the activity of the oxides of uranium and thorium is very similar.

Radio-activity of the Compounds of Uranium and Thorium.

The following are the figures I obtained with different uranium compounds. I have represented the intensity of the current in ampères by the letter i:—

$i \times 10^{11}$.

Metallic uranium (containing a little carbon)	2·3
Black oxide of uranium, U_2O_5	2·6
Green ,, ,, U_3O_4	1·8
Hydrated uranic acid	0·6
Uranate of sodium	1·2
,, potassium	1·2
,, ammonium	1·3
Uranium sulphate	0·7
Sulphate of uranium and potassium	0·7
Nitrate of uranium	0·7
Phosphate of copper and uranium	0·9
Oxysulphide of uranium	1·2

The thickness of the layer of the uranium compound used has little effect, provided that the layer is uniform. The following illustrate this point :—

	Thickness of layer. M.m.	$i \times 10^{11}$
Uranium oxide	0·5	2·7
,, ,,	3·0	3·0
Ammonium uranate	0·5	1·3
,, ,,	3·0	1·4

It may be concluded from this that the absorption of uranium rays by the substance which generates them is very great, since the rays proceeding from deep layers produce no significant effect.

The figures I obtained with thorium compounds enable me to state :—

Firstly, that the thickness of the layer used has considerable effect, especially in the case of the oxide.

Secondly, that the action is only regular if a sufficiently thin layer is used (e.g., 0·25 m.m.). On the contrary, when a thick layer of the substance is used (6 m.m.), the figures obtained vary between two extreme limits, especially in the case of the oxide :—

	Thickness of layer. M.m.	$i \times 10^{11}$
Thorium oxide	0·25	2·2
,, ,,	0·5	2·5
,, ,,	2·5	4·7
,, ,,	3·0	5·5 (mean)
,, ,,	6·0	5·5 ,,
Thorium sulphate	0·25	0·8 ,,

There is here some cause of irregularities which do not exist in the case of the uranium compounds. The figures obtained for a layer of oxide 6 m.m. thick varied between 3·7 and 7·3.

The experiments that I made on the absorption of uranium and thorium rays showed that those of thorium are more penetrating than those of uranium, and that the rays emitted by the oxide of thorium in a thick layer are more penetrating than those emitted by a thin layer of the same. The following figures (p. 13) give the fraction of the radiation transmitted by a sheet of aluminium 0·01 thick.

With the uranium compounds, the absorption is the same whatever be the compound used, which leads to the conclusion that the rays emitted by the different compounds are of the same nature.

Radio-active substance.	Fraction of radiation transmitted by the sheet.
Uranium	0·18
Uranium oxide, U_2O_5	0·20
Uranate of ammonium	0·20
Phosphate of uranium and copper	0·21
Thorium oxide of thickness 0·25 m.m	0·38
,, ,, 0·5 ,,	0·47
,, ,, 3·0 ,,	0·70
,, ,, 0·60 ,,	0·70
Thorium sulphate 0·25 ,,	0·38

The characteristics of the thorium radiation have formed the subject of very complete publications. Mr. Owens has demonstrated that a uniform current is only obtained after some time has elapsed, with an enclosed apparatus, and that the intensity of the current is greatly reduced under the influence of a current of air (which does not occur with the compounds of uranium). Mr. Rutherford has made similar experiments, and has explained them by the proposition that thorium and its compounds produce, besides the Becquerel rays, another *emanation*, composed of extremely minute particles, which remain radio-active for some time after their emission, and are capable of being swept along by a current of air.

The characteristics of the thorium radiation, which have reference to the thickness of the layer employed and to the action of air currents, have an intimate connection with the phenomenon of the *radio-activity induced, and of its propagation from place to place*. This phenomenon was observed for the first time with radium, and will be described later.

The radio-activity of thorium and uranium compounds appears as an *atomic property*. M. Becquerel has already observed that all uranium compounds are active, and had concluded that their activity was due to the presence of the element uranium; he also demonstrated that uranium was more active than its salts. I have investigated, from this point of view, the compounds of thorium and uranium, and have taken a great many measurements of their activity under different conditions. The result of all these determinations shows the radio-activity of these substances to be decidedly an atomic property. It seems to depend upon the presence of atoms of the two elements in question, and is not influenced by any change of physical state or chemical decomposition. The chemical combinations and mixtures containing uranium or thorium are active in proportion to the amount of the metal contained, all inactive material acting as inert bodies and absorbing the radiation.

Is Atomic Radio-activity a general Phenomenon?

As I have said above, I made experiments to discover whether substances other than compounds of uranium and thorium were radio-active. I undertook this research with the idea that it was scarcely probable that radio-activity, considered as an atomic property, should belong to a certain kind of matter to the exclusion of all other. The determinations I made permit me to say that, for chemical elements actually considered as such, including the rarest and most hypothetical, the compounds I investigated were always at least 100 times less active in my apparatus than metallic uranium.

The following is a summary of the substances experimented upon, either as the element or in combination:—

1. All the metals or non-metals easily procurable, and some, more rare, pure products obtained from the collection of M. Etard, at the Ecole de Physique et de Chimie Industrielles de la Ville de Paris.

2. The following rare bodies:—Gallium, germanium, neodymium, praseodymium, niobium, scandium, gadolinium, erbium, samarium, and rubidium (specimens lent by M. Demarçay), yttrium, ytterbium (lent by M. Urbain).

3. A large number of rocks and minerals.

Within the limits of sensitiveness of any apparatus, I found no simple substance, other than uranium and thorium, possessing atomic radio-activity. It will be suitable to add a few words here concerning phosphorus. White moist phosphorus, placed between the plates of the condenser, causes the air between the plates to conduct. However, I do not consider this body radio-active in the same manner as thorium and uranium. For, under these conditions, phosphorus becomes oxidised and emits luminous rays, whilst uranium and thorium compounds are radio-active without showing any chemical change which can be detected by any known means. Further, phosphorus is not active in the red variety, nor in a state of combination.

In a recent work, M. Bloch has demonstrated that phosphorus, undergoing oxidation in air, gives rise to slightly motile ions, which make the air conduct, and cause condensation of aqueous vapour.

Uranium and thorium are elements which possess the highest atomic weights (240 and 232); they occur frequently in the same minerals.

Radio-active Minerals.

I have examined many minerals in my apparatus; certain of them gave evidence of radio-activity, *e.g.*, pitchblende, thorite, orangite, fergusonite, cleveite, chalcolite, autunite, monazite, &c. The following is a table giving in ampères the intensity, *i*, of the current obtained with metallic uranium and with different minerals:—

$i \times 10^{11}$.

Uranium	2·3
Pitchblende from Johanngeorgenstadt	8·3
,, ,, Joachimsthal	7·0
,, ,, Pzibran	6·5
,, ,, Cornwallis	1·6
Cleveite	1·4
Chalcolite	5·2
Autunite	2·7
Various thorites	$\begin{cases} 0·1 \\ 0·3 \\ 0·7 \\ 1·3 \\ 1·4 \end{cases}$
Orangite	2·0
Monazite	0·5
Xenotime	0·03
Æschynite	0·7
Fergusonite (two samples)	$\begin{cases} 0·4 \\ 0·1 \end{cases}$
Samarskite	1·1
Niobite (two samples)	$\begin{cases} 0·1 \\ 0·3 \end{cases}$
Tantalite	0·02
Carnotite	6·2

The current obtained with orangite (native oxide of thorium) varied greatly with the thickness of the layer. By increasing this thickness from 0·25 m.m. to 6 m.m. the current increased from 1·8 to 2·3.

All the minerals which showed radio-activity contained uranium or thorium: their activity is therefore not surprising, but the intensity of the action in certain cases is unexpected. Thus pitchblendes (ores of uranium oxide) are found which are four times as active as metallic uranium. Chalcolite (double phosphate of copper and uranium) is twice as active as uranium. Autunite (phosphate of uranium and calcium) is as active as uranium. These facts do not accord with previous conclusions, according to which no mineral should be so active as thorium or uranium.

To throw light on this point, I prepared artificial chalcolite by the process of Debray, starting with the pure products. The process consists in mixing a solution of uranium nitrate with a solution of copper phosphate in phosphoric acid and warming to 50° or 60°. After some time, crystals of chalcolite appear in the liquid.

Chalcolite thus obtained possesses a perfectly normal activity, given by its composition; it is two and a-half times less active than uranium.

It therefore appeared probable that if pitchblende, chalcolite, and autunite possess so great a degree of activity, these substances contain a small quantity of a strongly radio-active body, differing from uranium and thorium and the simple bodies actually known. I thought that if this were indeed the case, I might hope to extract this substance from the ore by the ordinary methods of chemical analysis.

CHAPTER II.

Method of Research.

The results of the investigation of radio-active minerals, announced in the preceding chapter, led M. Curie and myself to endeavour to extract a new radio-active body from pitchblende. Our method of procedure could only be based on radio-activity, as we know of no other property of the hypothetical substance. The following is the method pursued for a research based on radio-activity:—The radio-activity of a compound is determined, and a chemical decomposition of this compound is effected; the radio-activity of all the products obtained is determined, having regard to the proportion in which the radio-active substance is distributed among them. In this way, an indication is obtained, which may to a certain extent be compared to that which spectrum analysis furnishes. In order to obtain comparable figures, the activity of the substances must be determined in the solid form well dried.

Polonium, Radium, Actinium.

The analysis of pitchblende with the help of the method just explained, led us to the discovery in this mineral of two strongly radio-active substances, chemically dissimilar:— Polonium, discovered by ourselves, and radium, which we discovered in conjunction with M. Bémont.

Polonium from the analytical point of view, is analogous to bismuth, and separates out with the latter. By one of the following methods of fractionating, bismuth products are obtained increasingly rich in polonium :—

1. Sublimation of the sulphides *in vacuo;* the active sulphide is much more volatile than bismuth sulphide.
2. Precipitation of solutions of the nitrate by water; the precipitate of the basic nitrate is much more active than the salt which remains in solution.
3. Precipitation by sulphuretted hydrogen of a hydrochloric acid solution, strongly acid; the precipitated sulphides are considerably more active than the salt which remains in solution.

Radium is a substance which accompanies the barium obtained from pitchblende; it resembles barium in its reactions, and is separated from it by difference of solubility of the chlorides in water, in dilute alcohol, or in water acidified with hydrochloric acid. We effect the separation of the chlorides of barium and radium by subjecting the mixture to fractional crystallisation, radium chloride being less soluble than that of barium.

A third strongly radio-active body has been identified in pitchblende by M. Debierne, who gave it the name of *actinium*. Actinium accompanies certain members of the iron group contained in pitchblende; it appears in particular allied to thorium, from which it has not yet been found possible to separate it. The extraction of actinium from pitchblende is a very difficult operation, the separations being as a rule incomplete.

All three of the new radio-active bodies occur in quite infinitesimal amount in pitchblende. In order to obtain them in a more concentrated condition, we were obliged to treat several tons of residue of the ore of uranium. The rough treatment was carried out in the factory; and this was followed by processes of purification and concentration. We thus succeeded in extracting from thousands of kilogrms. of crude material a few decigrammes of products which were exceedingly active as compared with the ore from which they were obtained. It is obvious that this process is long, arduous, and costly.

Other new radio-active bodies have been notified since the termination of our work. M. Giesel, on the one hand, and MM. Hoffmann and Strauss on the other, have announced the probable existence of a radio-active body similar to lead in its chemical properties. At present only a few samples of this substance have been obtained.

Radium is, so far, the only member of the new radio-active substances that has been isolated as the pure salt.

Spectrum of Radium.

It was of the first importance to check, by all possible means, the hypothesis, underlying this work, of new radio-active elements. In the case of radium, spectrum analysis was the means of confirming this hypothesis.

M. Demarçay undertook the examination of the new radio-active bodies by the searching methods which he employs in the study of photographic spark spectra.

The assistance of so competent a scientist was of the greatest value to us, and we are deeply grateful to him for having consented to take up this work. The results of the spectrum analysis brought conviction to us when we were still in doubt as to the interpretation of the results of our research.

The first specimens of fairly active barium chloride containing radium, examined by M. Demarçay, exhibited together with the barium lines a new line of considerable intensity and of wave-length $\lambda = 381\cdot47\ \mu\mu$ in the ultra-violet. With the more active products prepared subsequently, Demarçay saw the line $381\cdot47\ \mu\mu$ more distinctly; at the same time other new lines appeared, and the intensity of the new lines was comparable with that of the barium lines. A further concentration furnished a product for which the new spectrum predominated, and the three strongest barium lines, alone visible, merely indicated the presence of this metal as an impurity. This product may be looked upon as nearly pure radium chloride. Finally, by further purification, I obtained an exceedingly pure chloride, in the spectrum of which the two chief barium lines were scarcely visible.

The following is a list, according to Demarçay, of the principal radium lines for the portion of the spectrum included between $\lambda = 500\cdot0$ and $\lambda = 350\cdot0\ \mu\mu$. The intensity of each line is represented by a figure, the strongest being marked 16:—

λ.	Intensity.	λ.	Intensity.
482·63	10	460·03	3
472·69	5	453·35	9
469·98	3	443·61	8
469·21	7	434·06	12
468·30	14	381·47	16
464·19	4	364·96	12

All the lines are clear and narrow, the three lines 381·47,

468·30, 434·06 are strong, and equal the most intense of those actually known. Two well-marked misty bands are also visible in the spectrum. The first, which is symmetrical, extends from 463·10 to 462·19, with a maximum at 462·75. The second, which is stronger, fades towards the ultra-violet; it begins, sharply defined, at 446·37, and passes through a maximum at 445·52; the region of the maximum extends as far as 445·34, then a nebulous band, gradually fading, extends about as far as 439.

In the least refrangible part, not photographed in the spark spectrum, the only significant line is 566·5 (approx.), much more feeble, however, than 482·63.

The general aspect of the spectrum is that of the metals of the alkaline earths; these metals are known to have well-marked line spectra with certain nebulous bands.

According to Demarçay, the position of radium may be among the bodies possessing the most sensitive spectrum reaction. I also have concluded from the work of concentration, that in the first specimen examined, which showed clearly the line 3814·7, the proportion of radium must have been very small (perhaps about 0·02 per cent). Nevertheless, an activity fifty times as great as that of metallic uranium is required in order to distinguish clearly the principal radium line in the spectra photographed. With a sensitive electrometer the radio-activity of a substance only 1/100 of that of metallic uranium can be detected. It is clear that, in order to detect the presence of radium, the property of radio-activity is several thousand times more sensitive than the spectrum reaction.

Bismuth containing polonium and thorium containing actinium, both very active, examined by Demarçay, have so far each only yielded bismuth and thorium lines.

In a recent publication, M. Giesel, who is occupied in preparing radium, states that radium bromide gives a carmine flame colouration. The flame spectrum of radium contains two beautiful red bands, one line in the blue-green, and two faint lines in the violet.

Extraction of the New Radio-active Substances.

The first stage of the operation consists in extracting barium with radium from the ores of uranium, also bismuth with polonium and the rare earths containing actinium from the same. These three primary products having been obtained, the next step is in each case to endeavour to isolate the new radio-active body. This second part of the treatment consists of a process of fractionation. The difficulty of finding a very perfect means of separating closely allied elements is well known; methods of fractionation are therefore quite suitable. Besides this, when a mere trace of one element is mixed with another element, no method of complete separation could be applied to the mixture, even allowing that such a method was known; in fact, one would run the risk of losing the trace of the material to be separated.

The particular object of my work has been the isolation of radium and polonium. After working for several years, I have so far only succeeded in obtaining the former.

Pitchblende is an expensive ore, and we have given up the treatment of it in large quantities. In Europe the extraction of this ore is carried out in the mine of Joachimsthal, in Bohemia. The crushed ore is roasted with carbonate of soda, and the resulting material washed, first with warm water and then with dilute sulphuric acid. The solution contains the uranium, which gives pitchblende its value. The insoluble residue is rejected. This residue contains radio-active substances; its activity is four and a-half times that of metallic uranium. The Austrian Government, to whom the mine belongs, presented us with a ton of this residue for our research, and authorised the mine to give us several tons more of the material.

It was not very easy to apply the methods of the laboratory to the preliminary treatment of the residue in the factory. M. Debierne investigated this question, and organised the treatment in the factory. The most important point of his method is the conversion of the sulphates into carbonate by boiling the material with a concentrated solution of sodium carbonate. This method avoids the necessity of fusing with sodium carbonate.

The residue chiefly contains the sulphates of lead and calcium, silica, alumina, and iron oxide. In addition nearly all the metals are found in greater or smaller amount (copper, bismuth, zinc, cobalt, manganese, nickel, vanadium, antimony, thallium, rare earths, niobium, tantalum, arsenic, barium, &c.). Radium is found in this mixture as sulphate, and is the least soluble sulphate in it. In order to dissolve it, it is necessary to remove the sulphuric acid as far as possible. To do this, the residue is first treated with a boiling concentrated soda solution. The sulphuric acid combined with the lead, aluminium, and calcium passes, for the most part, into solution as sulphate of sodium, which is removed by repeatedly washing with water. The alkaline solution removes at the same time lead, silicon, and aluminium. The insoluble portion is attacked by ordinary hydrochloric acid. This operation completely disintegrates the material, and dissolves most of it. Polonium and actinium may be obtained from this solution; the former is precipitated by sulphuretted hydrogen, the latter is found in the hydrates precipitated by ammonia in the solution separated from the sulphides and oxidised. Radium remains in the insoluble portion. This portion is washed with water, and then treated with a boiling concentrated solution of carbonate of soda. This operation completes the transformation of the sulphates of barium and radium into carbonates. The material in then thoroughly washed with water, and then treated with dilute hydrochloric acid, quite free from sulphuric acid. The solution contains radium as well as polonium and actinium. It is filtered and precipitated with sulphuric acid. In this way the crude sulphates of barium containing radium and calcium, of lead, and of iron, and of a trace of actinium are obtained. The solution still contains a little actinium and polonium, which may be separated out as in the case of the first hydrochloric acid solution.

From one ton of residue 10 to 20 kilogrms. of crude sulphates are obtained, the activity of which is from thirty to sixty times as great as that of metallic uranium. They must now be purified. For this purpose they are boiled with sodium carbonate and transformed into the chlorides. The solution is treated with sulphuretted hydrogen, which gives a small quantity of active sulphides containing polonium. The solution is filtered, oxidised by means of chlorine, and precipitated with pure ammonia. The precipitated hydrates and oxides are very active, and the activity is due to actinium. The filtered solution is precipitated with sodium carbonate. The precipitated carbonates of the alkaline earths are washed and converted into chlorides. These chlorides are evaporated to dryness, and washed with pure concentrated hydrochloric acid. Calcium chloride dissolves almost entirely, whilst the chloride of barium and radium remains insoluble. Thus, from one ton of the original material about 8 kilogrms. of barium and radium chloride are obtained, of which the activity is about sixty times that of metallic uranium. The chloride is now ready for fractionation.

Polonium.

As I said above, by passing sulphuretted hydrogen through the various hydrochloric acid solutions obtained during the course of the process, active sulphides are precipitated, of which the activity is due to polonium. These sulphides chiefly contain bismuth, a little copper and lead; the latter metal occurs in relatively small amount, because it has been to a great extent removed by the soda solution, and because its chloride is only slightly soluble. Antimony and arsenic are found among the oxides only in the minutest quantity, their oxides having been dissolved by the soda. In order to obtain the very active sulphides, the following process was employed:—The solutions made strongly acid with hydrochloric acid were precipitated with sulphuretted hydrogen; the sulphides thus precipitated are very active, and are employed for the preparation of polonium; there remain in the solution substances not completely precipitated in presence of excess of hydrochloric acid (bismuth, lead, antimony). To complete the precipitation, the solution is diluted with water, and treated again with sulphuretted hydrogen, which gives a second precipitate of sulphides, much less active than the first, and which have generally been rejected. For the further purification of the sulphides, they are washed with ammonium sulphide, which removes the last remaining traces of antimony and arsenic. They are then washed with water and ammonium nitrate, and treated with dilute nitric acid. Complete solution never occurs; there is always an insoluble residue, more or less considerable, which can be treated afresh if it is judged expedient. The solution is reduced to a small volume and precipitated either by ammonia or by excess of water. In both cases the lead and the copper remain in solution; in the second case, a little bismuth, scarcely active at all, remains also in solution.

The precipitate of oxides or basic nitrates is subjected to fractionation in the following manner:—The precipitate is dissolved in nitric acid, and water is added to the solution until a sufficient quantity of precipitate is formed; it must be borne in mind that sometimes the precipitate does not at once appear. The precipitate is separated from the supernatant liquid, and re-dissolved in nitric acid, after which both the liquids thus obtained are re-precipitated with water, and treated as before. The different fractions are combined according to their activity, and concentration is carried out as far as possible. In this way is obtained a very small quantity of a substance of which the activity is very high, but which, nevertheless, has so far only shown bismuth lines in the spectroscope.

There is, unfortunately, little chance of obtaining the isolation of polonium by this means. The method of fractionation just described presents many difficulties, and the case is similar with other wet processes of fractionation. Whatever be the method employed, compounds are readily formed which are absolutely insoluble in dilute or concentrated acids. These compounds can only be redissolved by reducing them to the metallic state, *e.g.*, by fusion with potassium cyanide. Considering the number of operations necessary, this circumstance constitutes an enormous difficulty in the progress of the fractionation. This obstacle is the greater because polonium, once extracted from the pitchblende, diminishes in activity. This diminution of activity is slow, for a specimen of bismuth nitrate containing polonium only lost half its activity in eleven months.

No such difficulty occurs with radium. The radio-activity remains throughout an accurate guage of the concentration; the concentration itself presents no difficulty, and the progress of the work from the start can be constantly checked by spectral analysis.

When the phenomena of induced radio-activity, which will be discussed later on, were made known, it seemed obvious that polonium, which only shows the bismuth lines and whose activity diminishes with time, was not a new element, but bismuth made active by the vicinity of radium in the pitchblende. I am not sure that this opinion is correct. In the course of my prolonged work on polonium, I have noted chemical effects, which I have never observed either with ordinary bismuth or with bismuth made active by radium. These chemical effects are, in the first place, the extremely ready formation of insoluble compounds, of which I have spoken above (especially basic nitrates), and, in the second place, the colour and appearance of the precipitates obtained by adding water to the nitric acid solution of bismuth containing polonium. These precipitates are sometimes white, but more generally of a more or less vivid yellow, verging on red.

The absence of lines other than those of bismuth does not necessarily prove that the substance only contains bismuth, because bodies exist whose spectrum reaction is scarcely visible.

It would be necessary to prepare a small quantity of bismuth containing polonium in as concentrated a condition as possible, and to examine it chemically, in the first place determining the atomic weight of the metal. It has not yet been possible to carry out this research on account of the difficulties of a chemical nature already mentioned.

If polonium were proved to be a new element, it would be no less true that it cannot exist indefinitely in a strongly radio-active condition, at least when extracted from the ore. There are therefore two aspects of the question:— First, whether the activity of polonium is entirely induced by the proximity of substances themselves radio-active, in which case polonium would possess the faculty of acquiring atomic activity permanently, a faculty which does not appear to belong to any substance whatever; second, whether the activity of polonium is an inherent property, which is spontaneously destroyed under certain conditions, and persists under certain other conditions, such as those which exist in the ore. The phenomenon of atomic activity induced by contact is still so little understood, that we lack the ground on which to formulate any opinion on the matter.

(NOTE.—A work has recently appeared on polonium by M. Marckwald. He plunges a small rod of pure bismuth into a hydrochloric acid solution of the bismuth extracted from the pitchblende residue. After some time the rod becomes coated with a very active deposit, and the solution now contains only inactive bismuth. M. Marckwald also obtains a very active deposit by adding tin chloride to a hydrochloric acid solution of radio-active bismuth. From this he concludes that the active element is allied to tellurium, and gives it the name of *radiotellurium*. This active substance of M. Marckwald seems identical with polonium, from its behaviour, and from the easily absorbed rays it emits. The choice of a new name for this substance is futile in the present state of the question).

Preparation of the Pure Chloride of Radium.

The method by which I extracted pure radium chloride from barium chloride containing radium consists in first subjecting the mixture of the chlorides to fractional crystallisation in pure water, then in water to which hydrochloric acid has been added. The difference in solubility of the two chlorides is thus made use of, that of radium being less soluble than that of barium.

At the beginning of the fractionation, pure distilled water is used. The chloride is dissolved, and the solution raised to boiling-point, and allowed to crystallise by cooling in a covered capsule. Beautiful crystals form at the bottom, and the supernatant, saturated solution is easily decanted. If part of this solution be evaporated to dryness, the chloride obtained is found to be about five times less active than that which has crystallised out. The chloride is thus divided into two portions, A and B—portion A being more active than portion B. The operation is now repeated with each of the chlorides A and B, and in each case two new portions are obtained. When the crystallisation is finished, the less active fraction of chloride A is added to the more active fraction of chloride B, these two having approximately the same activity. Thus there are now three portions to undergo afresh the same treatment.

The number of portions is not allowed to increase indefinitely. The activity of the most soluble portion diminishes as the number increases. When its activity becomes inconsiderable, it is withdrawn from the fractionation. When the desired number of fractions has been obtained, fractionation of the least soluble portion is stopped (the richest in radium), and it is withdrawn from the remainder.

A fixed number of fractions is used in the process. After each series of operations, the saturated solution arising from one fraction is added to the crystals arising from the following fraction; but if after one of the series the most soluble fraction has been withdrawn, then, after the following series, a new fraction is made from the most soluble portion, and the crystals of the most active portion are withdrawn. By the successive alteration of these two processes, an extremely regular system of fractionation is obtained, in which the number of fractions and the activity of each remains constant, each being about five times as active as the subsequent one, and in which, on the one hand, an almost inactive product is removed, whilst, on the other, is obtained a chloride rich in radium. The amount of material contained in these fractions gradually diminishes, becoming less as the activity increases.

At first six fractions were used, and the activity of the chloride obtained at the end was only o·1 that of uranium.

When most of the inactive matter has been removed, and the fractions have become small, one fraction is removed from the one end, and another is added to the other end consisting of the active chloride previously removed. A chloride richer in radium than the preceding is thus obtained. This system is continued until the crystals obtained are pure radium chloride. If the fractionation has been thoroughly carried out, scarcely any trace of the intermediate products remain.

At an advanced stage of the fractionation, when the quantity of material in each fraction is small, the separation by crystallisation is less efficacious, the cooling being too rapid and the volume of the solution to be decanted too small. It is then advisable to add water containing a known quantity of hydrochloric acid; this quantity may be increased as the fractionation proceeds.

The advantage gained thus consists in increasing the quantity of the solution, the solubility of the chlorides being less in water acidified with hydrochloric acid than in pure water. By using water containing much acid, excellent separations are effected, and it is only necessary to work with three or four fractions.

The crystals, which form in very acid solution, are elongated needles, those of barium chloride having exactly the same appearance as those of radium chloride. Both show double refraction. Crystals of barium chloride containing radium are colourless, but when the proportion of radium becomes greater, they have a yellow colouration after some hours, verging on orange, and sometimes a beautiful pink. This colour disappears in solution. Crystals of pure radium chloride are not coloured, so that the colouration appears to be due to the mixture of radium and barium. The maximum colouration is obtained for a certain degree of radium present, and this fact serves to check the progress of the fractionation.

I have sometimes noticed the formation of a deposit composed of crystals of which one part remained uncoloured, whilst the other was coloured, and it seems possible that the colourless crystals might be sorted out.

The fractional precipitation of an aqueous solution of barium chloride by alcohol also leads to the isolation of radium chloride, which is the first to precipitate. This method, which I first employed, was finally abandoned for the one just described, which proceeds with more regularity. I have, however, occasionally made use of precipitation by alcohol to purify radium chloride which contains traces of barium chloride. The latter remains in the slightly aqueous alcoholic solution, and can thus be removed.

M. Giesel, who, since the publication of our first researches, has been preparing radio-active bodies, recommends the separation of barium and radium by fractional crystallisation in water from a mixture of the bromides. I can testify that this method is advantageous, especially in the first stages of the fractionation.

Determination of the Atomic Weight of Radium.

In the course of my work I determined at intervals the atomic weight of the metal contained in specimens of barium chloride containing radium. With each newly obtained product I carried the concentration as far as possible, so as to have from 0·1 grm. to 0·5 grm. of material containing most of the activity of the mixture. From this small quantity I precipitated with alcohol or with hydrochloric acid some milligrms. of chloride for spectral analysis. Thanks to his excellent method, Demarçay only required this small quantity of material to obtain the photograph of the spark spectrum. I made an atomic weight determination with the product remaining.

I employed the classic method of weighing as silver chloride the chlorine contained in a known weight of the anhydrous chloride. As control experiment, I determined the atomic weight of barium by the same method, under the same conditions, and with the same quantity of material, first 0·5 grm. and then 0·1 grm. The figures obtained were always between 137 and 138. I thus saw that the method gives satisfactory results, even with a very small quantity of material.

The first two determinations were made with chlorides, of which one was 230 times and the other 600 times as active as uranium. These two experiments gave the same figure as the experiment with the pure barium chloride. There was therefore no hope of finding a difference except by using a much more active product. The following experiment was made with a chloride, the activity of which was about 3500 times as great as that of uranium; and this experiment enabled me, for the first time, to observe a small but distinct difference; I found, as the mean atomic weight of the metal contained in this chloride, the number 140, which showed that the atomic weight of radium must be higher than that of barium. By using more and more active products, and obtaining spectra of radium of increasing intensity, I found that the figures obtained rose in proportion, as is seen in the following table (p. 28).

The figures of column A must only be looked upon as a rough estimate. The calculation of the activity of strongly radio-active bodies is difficult, for many reasons which will be discussed later.

A.	M.	
3500	140	Spectrum of radium faint.
4700	141	
7500	145·8	Spectrum of radium strong, but that of barium predominating.
Order of magnitude, 10⁶	173·8	The two spectra of almost equal intensity.
	225	Only a trace of barium present.

A represents the activity of the chloride, that of uranium being unity; M the atomic weight found.

At the termination of the processes described above, I obtained, in March, 1902, 0·12 grm. of radium chloride, of which Demarçay made the spectral analysis. This radium chloride, in the opinion of Demarçay, was fairly pure; its spectrum, however, showed the three principal barium lines with considerable intensity. I made four successive estimations of the chloride, the results of which are as follows:—

	Anhydrous radium chloride.	Silver chloride.	M.
I.	0·1150	0·1130	220·7
II.	0·1140	0·1119	223·0
III.	0·11135	0·1086	222·8
IV.	0·10925	0·10645	223·1

I then re-purified this chloride, and obtained a much purer substance, in the spectrum of which the two strongest barium lines were very faint. Given the sensitiveness of the spectrum reaction of barium, Demarçay estimated that the purified chloride contained only the merest traces of barium, incapable of influencing the atomic weight to an appreciable extent. I made three determinations with this perfectly pure radium chloride. The results were as follows:—

	Anhydrous radium chloride.	Silver chloride.	M.
I.	0·09192	0·08890	225·3
II.	0·08936	0·08627	225·8
III.	0·08839	0·08589	224·0

The mean of these numbers is 225. They were calculated in the same way as the preceding ones by considering radium as a bivalent element, the chloride having the formula $RaCl_2$, and taking for silver and chlorine the values $Ag = 107·8$, $Cl = 35·4$.

Hence the atomic weight of radium is $Ra = 225$.

The weighings were made with a Curie aperiodic balance, perfectly regulated, accurate to the twentieth of a milligrm. This direct reading balance permits of very rapid weighing, a condition which is essential in the case of the anhydrous chlorides of radium and barium, which gradually absorb moisture, in spite of the presence of desiccating substances in the balance. The bodies to be weighed were placed in a platinum crucible; this crucible had been long in use, and its weight did not vary the tenth part of a milligrm. during the course of one operation.

The hydrated chloride obtained by crystallisation was placed in the crucible and heated till converted into the anhydrous chloride. When the chloride has been kept for several hours at 100° its weight becomes constant, and does not change even if the temperature is raised to 200°. The anhydrous chloride thus obtained constitutes, therefore, a perfectly definite body.

The following is a series of determinations on this point. The chloride (100 m.g.) is dried in the oven at 55°, and placed in a desiccator over anhydrous phosphoric acid; it then gradually loses weight, which proves that it still contains moisture; in the course of twelve hours the loss was 3 m.g. The chloride is replaced in the stove, and the temperature raised to 100°. During this process, the chloride lost 6·3 m.g. in weight. After being left three hours fifteen minutes in the oven, it lost 2·5 m.g. more. The temperature was maintained for forty-five minutes between 100° and 120°, which caused a loss of weight of 0·1 m.g. Then after being kept for thirty minutes at 125°, the chloride showed no diminution in weight. Then, however, after thirty minutes at 150°, it lost 0·1 m.g. Finally, after being heated for four hours at 200°, it lost 0·15 m.g. During these operations the crucible varied from 0·05 m.g.

After each determination of the atomic weight, the radium was converted into the chloride in the following manner:—To the solution containing the weighed radium nitrate and excess of silver nitrate was added pure hydrochloric acid; the silver chloride was filtered off; the solution was evaporated to dryness several times with excess of pure hydrochloric acid. In this way the nitric acid is entirely removed.

The precipitated silver chloride was always radio-active and phosphorescent. In determining the amount of silver contained in it, I satisfied myself that no ponderable amount of radium had been carried down with it out of the solution. The method I pursued was to reduce the silver chloride precipitated in the crucible by hydrogen generated from dilute hydrochloric acid and zinc; after washing, the crucible was weighed with the metallic silver contained in it.

I made another experiment which showed that the weight of radium chloride regenerated was the same as that before beginning the operation.

These verifications are not so reliable as direct experiments; but they serve to indicate the absence of any significant error.

From its chemical properties, radium is an element of the group of alkaline earths, being the member next above barium.

From its atomic weight also, radium takes its place in Mendeleeff's table after barium with the alkaline earth metals, in the row which already contains uranium and thorium.

Characteristics of the Radium Salts.

The salts of radium, chloride, nitrate, carbonate, and sulphate, resemble those of barium, when freshly prepared, but they gradually become coloured.

All the radium salts are luminous in the dark.

In their chemical properties, the salts of radium are absolutely analogous to the corresponding salts of barium. However, radium chloride is less soluble than barium chloride; the solubility of the nitrates in water is approximately the same.

The salts of radium are the source of a spontaneous and continuous evolution of heat.

Fractionation of Ordinary Barium Chloride.

We have endeavoured to determine whether commercial barium chloride contains small quantities of radium chloride, which escape detection with the means of estimation at our command. For this purpose we fractionated a great quantity of commercial barium chloride, in the hope of thus concentrating the trace of radium chloride if such were present.

Fifty kilos. of commercial barium chloride were dissolved in water; the solution was precipitated by hydrochloric acid free from sulphuric acid, which yielded 20 kilos. of the precipitated chloride. This was dissolved in water and partially precipitated by hydrochloric acid, which gave 8·5 kilos. of precipitated chloride. This chloride was fractionated by the method used for the barium chloride containing radium; and at the end of the process, 10 grms. of chloride were obtained, corresponding to the least soluble part. This chloride showed no radio-activity; it therefore contained no radium; this substance is, consequently, absent from the ores of barium.

CHAPTER III.
Radiation of the New Radio-active Substances.
Methods of Investigation of the Radiation.

In order to investigate the radiation emitted by radio-active bodies, any one of the properties of this radiation can be utilised. Thus the action of the rays on photographic plates may serve, or their property of ionisation of the air, which renders it a conductor, or their capacity for causing fluorescence of certain bodies. Henceforth, in speaking of these different methods of working, I shall use the expressions radiographic method, electrical method, fluoroscopic method.

The first two have been used from the beginning in the study of uranium rays; the fluoroscopic method can only be applied in the case of the new bodies which are strongly radio-active, for the feebly active bodies such as uranium and thorium produce no appreciable fluorescence. The electrical method is the only one which serves for exact determinations of intensity; the other two are specially adapted for giving qualitative results, and only furnish rough approximations. The results obtained with the three methods just considered are not strictly comparable the one with the other. The sensitive plate, the gas which is ionised, the fluorescent screen, are in reality receivers, which absorb the energy of the radiation, and transform it into another kind of energy, chemical energy, ionic energy, or luminous energy. Each receiver absorbs a fraction of the radiation, which depends essentially upon its nature. Later on, we shall see that the radiation is complex, that the fractions of the radiation absorbed by the different receivers may differ among themselves both quantitatively and qualitatively. Finally, it is neither evident, nor even probable, that the energy absorbed is entirely transformed by the receiver into the form that we wish for observation; part of this energy may be transformed into heat, into the evolution of secondary radiations which may or may not assist in the production of the observed phenomenon, into chemical action which differs from that under observation, &c., and here also the effective action of the receiver, with reference to the end we have in view, depends essentially upon the nature of that receiver.

Let us compare two radio active substances, one containing radium and the other polonium, and which show an equal degree of activity in the condenser of Fig. 1. If each is covered with a thin leaf of aluminium, the second appears considerably less active than the first, and the same is the case when they are placed under the same fluorescent screen, if the latter is of sufficient thickness, or is placed at a certain distance from the two radio-active bodies.

Energy of Radiation.

Whatever be the method of research employed, the energy of radiation of the new radio-active substances is always found to be considerably greater than that of uranium and thorium. Thus it is that, at a short distance, they act instantaneously upon a photographic plate, whereas an exposure of twenty-four hours is necessary when operating with uranium and thorium. A fluorescent screen is vividly illuminated by contact with the new radio-active bodies, whilst no trace of luminosity is visible with uranium and thorium. Finally, the ionising action upon air is considerably stronger in the ratio of 10^6 approximately. But it is, strictly speaking, not possible to estimate the *total intensity of the radiation*, as in the case of uranium, by the electrical method described at the beginning (Fig. 1). With uranium, for example, the radiation is almost completely absorbed by the layer of air between the plates, and the limiting current is reached at a tension of 100 volts. But the case is different for strongly radio-active bodies. One portion of the radiation of radium consists of very penetrating rays, which penetrate the condenser and the metallic plates, and are not utilised in ionising the air between the plates. Further, the limiting current cannot always be obtained for the tensions supplied; for example, with very active polonium the current remains proportional to the tension between 100 and 500 volts. Therefore the experimental conditions which give a simple interpretation are not realised, and, consequently, the numbers obtained cannot be taken as representing the measurement of the total radiation; they merely point to a rough approximation.

Complex Nature of the Radiation.

The researches of various physicists (MM. Becquerel, Meyer and von Schweidler, Giesel, Villard, Rutherford, M. and Mdme. Curie) have proved the complex nature of the radiation of radio-active bodies. It will be convenient to specify three kinds of rays, which I shall denote, according to the notation adopted by Mr. Rutherford, by the letters α, β, γ.

I. The α-rays are very slightly penetrating, and appear to constitute the principal part of the radiation. These rays are characterised by the laws by which they are absorbed by matter. The magnetic field acts very slightly upon them, and they were formerly thought to be quite unaffected by the action of this field. However, in a strong magnetic field, the α-rays are slightly deflected; the deflection is caused in the same manner as with cathode rays, but the direction of the deflection is reversed; it is the same as for the canal rays of the Crookes tubes.

II. The β-rays are less absorbable as a whole than the preceding ones. They are deflected by a magnetic field in the same manner and direction as cathode rays.

III. The γ-rays are penetrating rays, unaffected by the magnetic field, and comparable to Röntgen rays.

Consider the following imaginary experiment :—Some radium, R, is placed at the bottom of a small deep cavity, hollowed in a block of lead, P (Fig. 4). A sheaf of rays, rectilinear and slightly expanded, streams from the receptacle. Let us suppose that a strong uniform magnetic field is established in the neighbourhood of the receptacle, normal to the plane of the figure and directed towards the back.

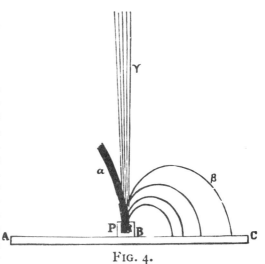

FIG. 4.

The three groups of rays, α, β, γ, will now be separated. Then rather faint γ-rays continue in their straight path without a trace of deviation. The β-rays are deflected in the manner of cathode rays, and describe circular paths in the plane of the figure. If the receptacle is placed on a photographic plate, A C, the portion, B C, of the plate which receives the β-rays is acted upon. Lastly, the α-rays form a very intense shaft which is slightly deflected, and which is soon absorbed by the air. These rays describe in the plane of the figure a path of great curvature, the direction of the deflection being the reverse of that with the β-rays.

If the receptacle is covered with a thin sheet of aluminium (0·1 m.m. thick), the α-rays are suppressed almost entirely, the β-rays are lessened, and the γ-rays do not appear to be absorbed to any great extent.

Action of the Magnetic Field.

We have seen that the rays emitted by radio-active bodies have many properties common to cathode rays and to Röntgen rays. Cathode rays, as well as Röntgen rays, ionise the air, act on photographic plates, cause fluorescence, undergo no regular deflection. But the cathode rays differ from Röntgen rays in being deflected from their rectilinear path by the action of the magnetic field, and in the transportation of charges of negative electricity.

The fact that the magnetic field acts upon the rays emitted by radio-active substances was discovered almost simultaneously by MM. Giesel, Meyer and von Schweidler, and Becquerel. These physicists observed that the rays of radio-active substances are deflected by the magnetic field in the same manner and direction as the cathode rays; their observations were in relation to the β-rays.

M. Curie demonstrated that the radiation of radium comprises two groups of quite distinct rays, of which one is readily deflected by the magnetic field (β-rays), whilst the other seems to be unaffected by the action of this field (α- and γ-rays.

M. Becquerel did not find that the specimens of polonium prepared by us emitted rays of the cathode kind. On the contrary, he first noticed the effect of the magnetic field on a specimen of polonium prepared by himself. None of the polonium prepared by us ever gave rise to rays of the cathode order.

The polonium of M. Giesel only gives rise to these rays when recently prepared, and it is probable that the emission is due to the phenomenon of induced radio-activity of which we shall speak later.

The following are experiments which prove that one portion of the radiation of radium, and one portion only, consists of easily deflected rays (β-rays). These experiments were done according to the electrical method.

The radio-active body A (Fig. 5) sends forth radiations in the direction A D between the plates P and P'. The plate P is now at a potential of 500 volts, plate P' is connected to an electrometer and to a quartz electric piezometer. The intensity of the current passing through the air under the influence of the radiations is measured. The magnetic field can be established at will perpendicular to the plane of the figure over the whole region E E E E. If the rays are deflected, even slightly, they no longer pass between the plates, and the current is suppressed. The region of the passage of the rays is surrounded with masses of lead, B, B', B", and by the armatures of the electro-magnet; when the rays are deflected, they are absorbed by the masses of lead B and B'.

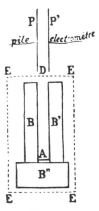

FIG. 5.

The results obtained depend essentially on the distance, A D, of the radiating substance, A, from the condenser at D. If the distance A D is great enough (greater than 7 c.m.), most of the radium rays (90 to 100 per cent) arriving at the condenser are deflected and suppressed for a field of 2500 units. These are the β-rays. If the distance A D is less than 65 m.m., a smaller part of the rays are deflected by the action of the field; this portion is also entirely deflected by a field of 2500 units, and the proportion of the rays suppressed is not increased by increasing the field from 2500 to 7000 units.

The proportion of the rays not suppressed by the field increases with decrease of the distance, A D, between the radiating body and the condenser. For small distances, the rays which can be easily deflected form a very small fraction of the total radiation. The penetrating rays are therefore, for the most part, deviable rays of the cathode order (β-rays).

Under the experimental conditions just described, the action of the magnetic field on the α-rays could not be well observed for the fields employed. The chief radiation, apparently undergoing no deflection, observed at a short distance from the radiating source, consisted of α-rays; the undeflected radiation observed at a greater distance consisted of γ-rays.

If an absorbing lamina (aluminium or black paper) is placed in the path of the bundle of rays, those which pass through are nearly all deflected by the field in such a way that, with the aid of the screen and the magnetic field, almost all the radiation is suppressed in the condenser, the remainder being due to the γ-rays, the proportion of which is small. The α-rays are absorbed by the screen.

An aluminium plate of 1/100 m.m. thickness is sufficient for the suppression of almost all the rays not readily deflected when the substance is far enough from the condenser; for smaller distances (34 m.m. and 51 m.m.) two pieces of this aluminium foil are necessary to give the same result.

Similar determinations were made with four substances containing radium (chlorides or carbonates) of very different activity; analogous results were obtained.

It may be remarked that, in all cases, the penetrating rays deflected by the magnet (β-rays) form only a small fraction of the total radiation; they influence but slightly the determinations in which the whole radiation is made use of to produce conductivity of the air.

The radiation emitted by polonium may be studied by the electrical method. When the distance, A D, of the polonium from the condenser is varied, no current is observed at first while the distance is fairly great; on nearing the polonium, the radiation suddenly becomes manifest with great intensity; the current then increases uniformly whilst approaching the polonium, but the magnetic field produces no appreciable effect under these conditions. The radiation of polonium is apparently limited in space, and does not pass into the air beyond a kind of sheath surrounding the substance to a thickness of several centimetres.

The interpretation of the experiments I have just described must be accompanied by some important general reservations. In speaking of the proportion of the rays deflected by the magnet, I refer only to that portion of the radiation capable of causing a current in the condenser. In employing the fluorescent action of the Becquerel rays, or their action on photographic plates, the proportion would probably be different—a measure of intensity having, as a rule, no meaning except for the method of measurement adopted.

The rays of polonium are α-rays. In the experiments just described, I observed no action of the magnetic field upon them, but the experimental conditions were such that a slight deflection would pass unnoticed.

The experiments made by the radiographic method confirmed the preceding results. Taking radium as the source of radiation, and receiving the impression on a plate parallel to the primitive shaft and normal to the field, a very clear print is obtained of two shafts separated by the action of the field, the one deflected, the other not deflected. The β-rays constitute the deflected beam; the α-rays, being very slightly deflected, are not to be distinguished from the undeflected bundle of the γ-rays.

Deflected β-Rays.

The experiments of M. Giesel and MM. Meyer and von Schweidler showed that the radiation of the radio-active bodies is, in part at least, deflected by a magnetic field, and that this deflection resembles that of the cathode rays. M. Becquerel investigated the action of the field on the rays by the radiographic method. The experimental arrangement was that of Fig. 4. The radium was placed in the lead receptacle, P, and this receptacle was placed on the sensitive face of a photographic plate, A C, covered with black paper. The whole was placed between the poles of an electro-magnet, the magnetic field being normal to the plane of the figure.

If the field is directed to the back of this plane, the part B C of the plate is acted upon by rays which, after having described circular paths, return to the plate and strike it at a right angle. These rays are β-rays.

M. Becquerel has demonstrated that the impression consists of a wide diffused band, a continuous spectrum indeed, showing that the sheaf of deviable rays emitted by the source is formed of an infinite number of radiations unequally deflected. If the gelatin of the plate be covered with different absorbent screens (paper, glass, metals), one portion of the spectrum is suppressed, and it is found that the rays most deflected by the magnetic field—otherwise those which have the smallest radius of curvature—are the most completely absorbed. With each screen, the impression on the plate begins at a certain distance from the source of radiation, this distance being proportional to the absorptive power of the screen.

Charge of the Deflected Rays.

The cathode rays are, as shown by M. Perrin, charged with negative electricity. Further, according to the experiments of M. Perrin and M. Lenard, they are capable of carrying their charge through the metallic envelopes connected to earth and through isolating screens. At every point where the cathode rays are absorbed, there is a continuous evolution of negative electricity. We have proved that the same is the case for the deflected β-rays of radium. *The deviable β-rays of radium are charged with negative electricity.*

(NOTE.—Let the radio-active substance be placed on one of the plates of a condenser, this plate being connected to earth ; the second plate is connected to an electrometer, it receives and absorbs the rays emitted by the substance. If the rays are charged, a continuous flow of electricity into the electrometer should be observed. In this experiment, carried out in air, we were not able to detect a charge accompanying the rays, but such an experiment is not delicate. The air between the plates being caused by the rays to conduct, the electrometer is no longer isolated, and can only respond to charges if these be sufficiently strong. In order that the α-rays may not interfere with the experiment, they may be suppressed by covering the source of radiation with a thin metallic screen. We repeated this experiment, without more success, by causing the rays to pass through the interior of a Faraday cylinder in connection with the electrometer).

According to the preceding experiments, it was evident that the charge of the rays of the radiating body employed was a weak one.

In order to fix a feeble evolution of electricity upon the conductor which absorbs the rays, this conductor should be completely insulated ; this is effected by screening it from the air, either by placing it in a tube with a very perfect vacuum, or by surrounding it with a good solid dielectric. We employed the latter arrangement.

A conducting disc, M M (Fig. 6), is connected by the wire, t, to the electrometer ; the disc and wire are completely enveloped by the insulating substance $i\ i\ i\ i$; the whole is again surrounded with the metallic covering, E E E E, which is in electric connection with the earth. The insulator, $p\ p$, and the metallic envelope are very thin upon one of the faces of the disc. This face is exposed to the radiation of the barium and radium salt, R, placed outside in a lead receptacle. The rays emitted by the radium penetrate the metallic envelope and the insulating lamina, $p\ p$, and are absorbed by the metallic disc, M M. The latter

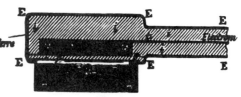

FIG. 6.

then becomes the source of a continuous evolution of negative electricity, as determined by the electrometer, and is measured by means of a quartz piezometer.

The current thus created is very weak. With very active barium-radium chloride, forming a layer of 2·5 sq. c.m. in area, and of 0·2 c.m. in thickness, a current of magnitude 10^{-11} ampères is obtained, the rays utilised having traversed, before being absorbed by the disc M M, a thickness of aluminium of 0·01 m.m., and a thickness of ebonite of 0·3 m.m.

We used successively lead, copper, and zinc for the disc M M, ebonite and paraffin for the insulator; the results obtained were the same.

The current diminishes with increasing distance from the source of radiation, R, also when a less active product is used.

We obtained the same results again when the disc M M is replaced by a Faraday cylinder filled with air, and covered outside with insulating material. The opening of the cylinder, closed by the thin insulating plate, $p\,p$, was opposite the radiating source.

Finally, we made the inverse experiment, which was to place the lead receptacle with the radium in the centre of

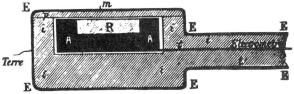

FIG. 7.

the insulating material and in connection with the electrometer (Fig. 7), the whole being surrounded with the metallic covering connected to earth.

Under these conditions, it is evident from the electrometer that the radium has a positive charge equal in magnitude to the negative charge of the former experiment. The radium rays penetrate the thin dielectric plate, $p\,p$, and leave the conductor inside carrying with them negative electricity.

The α-rays of radium do not interfere in these experiments, being almost completely absorbed by a very thin layer of matter. The method just described is not suitable for the study of the charge of the rays of polonium, these rays very slightly penetrating. We observed no indication of any charge in the case of polonium, which gives rise to α-rays only; but, for the reason just given, no conclusion can be drawn from this.

Thus, in the case of the deflected β-rays of radium, as in the case of cathode rays, the rays carry a charge of electricity. But, hitherto, the existence of electric charges uncombined with matter has been unknown. In the study of the emission of the β-rays of radium, we are therefore led to make use of the theory which is in vogue for the study of cathode rays. In this ballistic theory, formulated by Sir William Crookes, since developed and completed by Prof. J. J. Thomson, the cathode rays consist of extremely minute particles, which are hurled from the cathode with great velocity, and which are charged with negative electricity. We might similarly conceive that radium sends into space negatively electrified particles.

A specimen of radium, enclosed in a solid thin perfectly insulated envelope, should become spontaneously charged to a very high potential. By the ballistic hypothesis the potential would increase until the potential difference of the surrounding conductors became sufficient to hinder the ejection of the electrified particles and to cause their return to the source of radiation.

We have performed an experiment on these lines. A specimen of very active radium was enclosed for some time in a glass vessel. In order to open the vessel, we made a trace on the glass with a glass cutter. Whilst so doing, we clearly heard the report of a spark, and upon examining the vessel with a magnifying glass, we observed that the glass had been pierced by a spark at the spot where it had been weakened by the scratch. The phenomenon produced is comparable to the rupture of the glass of an overcharged Leyden jar.

The same phenomenon occurred with another glass. Further, at the moment of the passing of the spark, M. Curie, who was holding the glass, felt the electric shock of discharge in his fingers.

Certain kinds of glass have good insulating properties. If the radium is enclosed in a sealed glass vessel, well insulated, it is to be expected that, at a given moment, the vessel will be spontaneously perforated.

Radium is the first example of a body which is spontaneously charged with electricity.

Action of the Electric Field upon the Deflected β-Rays of Radium.

The β-rays of radium, being analogous to the cathode rays, should be deflected by an electric field in a manner similar to the latter; *i.e.*, as would a particle of matter negatively charged and hurled into space with a great velocity. The existence of such a deflection has been demonstrated both by M. Dorn and M. Becquerel.

Let us consider the case of a ray which traverses the space situated between the two plates of a condenser. Suppose the direction of the ray parallel to the plates: when an electric field is established between the latter, the ray is subjected to the action of this uniform field along its whole path in the condenser l. By reason of this action the ray is deflected towards the positive plate and describes the arc of a parabola ; on leaving the field, it continues its path in a straight line, following the tangent to the arc of the parabola at the point of exit. The ray can be received on a photographic plate perpendicular to its original direction. Observations are taken of the impression produced on the plate when the field is zero, and when it has a known value, and from that is deduced the value of the deflection, δ, which is the distance of the points in which the new direction of the ray and its original direction meet a common plane perpendicular to the original direction. If h is the distance of this plane from the condenser, $i.e.$, at the edge of the field, we have, by a simple calculation,—

$$\delta = \frac{e F l \left(\frac{l}{2} + h\right)}{mv^2};$$

m being the mass of the moving particles, e its charge, v its velocity, and F the strength of the field.

The experiments of M. Becquerel enable him to assign a value approaching to δ.

Relation of the Charge to the Mass for a Particle Negatively Charged Emitted by Radium.

When a material particle having a mass m and a negative charge e, is projected with a velocity v into a uniform magnetic field perpendicular to its initial velocity, this particle describes, in a plane normal to the field and passing through its initial velocity, an arc of a circle of radius ρ, so that—H being the strength of the field—we have the relation—

$$H\rho = \frac{m}{e} v.$$

If, for the same ray, the deflection, δ, and the radius of curvature, ρ, be measured in a magnetic field, values could be found from these two experiments for the ratio $\frac{e}{m}$ and for the velocity, v.

The experiments of M. Becquerel threw the first light upon this subject. They gave for the ratio $\frac{e}{m}$ a value approximately equal to 10^7 absolute electro-magnetic units, and for v a magnitude of $1\cdot 6 \times 10^{10}$. These values are of the same order of magnitude as those of the cathode rays.

Accurate experiments have been made on the same subject by M. Kaufmann. This physicist subjected a narrow beam of radium rays to the simultaneous action of an electric field and a magnetic field, the two fields being uniform and having a similar direction, normal to the original direction of the beam. The impression produced on a plate normal to the primitive beam and placed beyond the limits of the field with reference to the source, has the form of a curve, each point of which corresponds to one of the original beam. The most penetrating and least deflected rays are at the same time those with the greatest velocity.

It follows from the experiments of M. Kaufmann, that for the radium rays, of which the velocity is considerably greater than that of the cathode rays, the ratio $\frac{e}{m}$ decreases, while the velocity increases.

According to the researches of J. J. Thomson and Townsend, we may assume that the moving particle, which constitutes the ray, possesses a charge, e, equal to that carried by an atom of hydrogen during electrolysis, this charge being the same for all the rays. We are therefore led to the conclusion that the mass of the particle, m, increases with increase of velocity.

These theoretical considerations lead to the idea that the inertia of the particle is due to its state of charge during motion, the velocity of an electric charge in motion being incapable of modification without expenditure of energy. To state it otherwise, the inertia of the particle is of electro-magnetic origin, and the mass of the particle is—in part at least—a virtual mass or an electro-magnetic mass. M. Abraham goes further, and assumes that the mass of the particle is entirely an electro-magnetic mass. If, according to this hypothesis, the value of this mass, m, be calculated for a known velocity, v, we find that m approaches infinity when v approaches the velocity of light, and that m approaches a constant value when the velocity, v, is much less than that of light. The experiments of M. Kaufmann are in agreement with the results of this theory, the importance of which is great because it foreshadows the possibility of establishing mechanical bases upon the dynamical of little particles of matter charged in a state of motion.

These are the figures obtained by M. Kaufmann for $\dfrac{e}{m}$ and v.

$\dfrac{e}{m}$ Electro-magnetic units.	$v\,\dfrac{\text{c.m.}}{\text{sec.}}$	
$1\cdot865 \times 10^7$	$0\cdot7 \times 10^{10}$	For cathode rays (Simon).
$1\cdot31$,,	$2\cdot36$,,	⎫
$1\cdot17$,,	$2\cdot48$,,	⎪
$0\cdot97$,,	$2\cdot59$,,	⎬ For radium rays (Kaufmann).
$0\cdot77$,,	$2\cdot72$,,	⎪
$0\cdot63$,,	$2\cdot83$,,	⎭

M. Kaufmann concludes, from comparison of his experiments with the theory, that the limiting value of the ratio $\dfrac{e}{m}$ for radium rays of relatively small velocity would be the same as the value $\dfrac{e}{m}$ for cathode rays.

The most complete experiments of M. Kaufmann were made with a minute quantity of pure radium chloride, with which we provided him.

According to M. Kaufmann's experiments, certain β-rays of radium possess a velocity very near to that of light. These rapid rays seem to possess great penetrating capacity towards matter.

Action of the Magnetic Field upon the α-Rays.

In a recent work, Mr. Rutherford announced that, in a powerful electric or magnetic field, the α-rays of radium are slightly deflected, in the manner of particles positively electrified and possessing great velocity. Mr. Rutherford concludes from his experiments that the velocity of the α rays is of the order of magnitude $2\cdot5 \times 10^9\,\dfrac{\text{c.m.}}{\text{sec.}}$ and that the ratio $\dfrac{e}{m}$ for these rays is of the order of magnitude 6×10^3, which is 10^4 times as great as for the deflected β-rays. We shall see later that these conclusions of Mr. Rutherford are in agreement with the properties already known of the α-radiation, and that they account, in part at least, for the law of absorption of this radiation.

The experiments of Mr. Rutherford have been confirmed by M. Becquerel. M. Becquerel has further demonstrated that polonium rays behave in a magnetic field like the α-rays of radium, and that, for the same field, they seem to have the same curvature as the latter.

It also appears from M. Becquerel's experiments that the α-rays do not form a magnetic spectrum, but act rather like a homogeneous radiation, all the rays being equally deflected.

Action of the Magnetic Field on the Rays of other Radio-active Substances.

We have just seen that radium gives off α-rays comparable to the tube rays, β-rays comparable to cathode rays, and γ-rays which are penetrating and not deflected. Polonium gives off α-rays only. Amongst the other radio-active substances, actinium seems to behave like radium, but the study of its radiation has not yet advanced so far as in the case of radium. As regards the faintly radio-active bodies, we know to-day that uranium and thorium give rise to α-rays as well as β-rays (Becquerel, Rutherford).

Proportion of β-Rays in the Radiation of Radium.

As I have already mentioned, the proportion of β-rays increases with increase of distance from the source of radiation. These rays never occur alone, and for great distances the presence of γ-rays is always discernible. The presence of very penetrating, undeflected rays in the radiation of radium was first observed by M. Villard. These rays constitute only a small portion of the radiation measured by the electrical method, and their presence escaped our notice in our first experiments, so that we believed falsely that the radiation at great distances contained only rays capable of deflection.

The following are the numerical results obtained with experiments made by the electrical method with an apparatus similar to that of Fig. 5. The radium was only separated from the condenser by the surrounding air. I shall indicate by the letter d the distance from the source of radiation to the condenser. The numbers of the second line represent the current subsisting when the magnetic field is acting, supposing the current obtained with no field equal to 100 for each distance. These numbers may be considered as giving the percentage of the total α- and γ-rays, the deflection of the α-rays having been scarcely observable with the conditions employed.

At great distances there are no α-rays, and the undeflected radiation is therefore of the γ kind only.

Experiments made at short distances:—

d, in centimetres	3·4	5·1	6·0	6·5
Percentage of undeflected rays	74	56	33	11

Experiments made at long distances with a product considerably more active than that which was used for the preceding series:—

d, in centimetres	14	30	53	80	98
Percentage of undeflected rays	12	14	17	14	16
d	124	157			
Percentage of undeflected rays	14	11			

It is thus evident that after a certain distance the proportion of undeflected rays in the radiation is approximately constant. These rays probably all belong to the γ species.

The following is another series of experiments in which the radium was enclosed in a very narrow glass tube, placed below the condenser and parallel to the plates. The rays emitted traversed a certain thickness of glass and air before entering the condenser:—

d, in centimetres	2·5	3·3	4·1	5·9	7·5	9·6	11·3
Percentage rays not deflected	33	33	21	16	14	10	9
d	13·9	17·2					
Percentage rays not deflected	9	10					

As in the preceding experiments, the number of the second line approximate to a constant value, when the distance d increases, but the limit is reached for smaller distances than in the preceding series, because the a-rays have been more completely absorbed by the glass than the β- and γ-rays.

The following experiment shows that a thin sheet of aluminium (0·01 m.m. thick) absorbs principally a-rays. The product being placed 5 c.m. from the condenser, the proportion of rays other than β, when the magnetic field is acting, is about 71 per cent. When the same substance is covered with the sheet of aluminium, the distance remaining the same, the radiation transmitted is found to be almost totally deflected by the magnetic field, the a-rays having been absorbed by the aluminium. The same result is obtained when paper is used as the absorbing screen.

The greatest part of the radiation of radium consists of a-rays, which are probably emitted principally by the superficial layer of the radiating matter. When the thickness of the layer of radiating matter is varied, the intensity of the current increases with this thickness; the increase is not proportional to the thickness for the whole of the radiation; it is, moreover, more considerable for the β-rays than for the a-rays, so that the proportion of β-rays increases with the thickness of the active layer. The source of radiation being placed at a distance of 5 c.m. from the condenser, it is found that for a thickness equal to 0·4 m.m. of the active layer, the total radiation is given by the number 28, and the proportion of the β-rays is 29 per cent. By making the layer 2 m.m. thick, *i.e.*, five times as thick, a total radiation equal to 102, and a proportion of β-rays equal to 45 per cent are obtained. The total radiation which exists at this distance has therefore been increased in the ratio of 3·6, and the β-radiation has become five times as strong.

The preceding experiments were made by the electrical method. When the radiographic method is used, certain results seem to be in contradiction with what precedes. In the experiments of M. Villard, a beam of radium rays, subjected to the action of the magnetic field, was received on to a pile of photographic plates. The undeflected and penetrating γ-beam passed through all the plates, leaving its trace on each. The deflected β-beam produced an impression on the first plate only. This beam appeared therefore to contain no rays of great penetration.

On the contrary, in our experiments a beam which is propagated in the air contains at the greatest distances accessible to observation about 9/10 of β-rays, and the same is the case when the source of radiation is enclosed in a little sealed glass vessel. In M. Villard's experiments, these deflected and penetrating β-rays did not affect the photographic plates beyond the first, because they are to a great extent diffused in all directions by the first solid obstacle encountered, and no longer form a beam. In our experiments the rays given off by radium and transmitted through the glass of the vessel were also probably scattered by the glass, but the vessel being very small would itself act as a source of β-rays at its surface, and we were able to follow the course of the latter to a great distance from the vessel.

The cathode rays of Crookes tubes can only traverse very thin screens (aluminium screens of 0·01 m.m. thickness). A beam of rays striking the screen normally is scattered in all directions; but the diffusion becomes less with diminishing thickness of the screen, and for very thin screens the emerging beam is practically the prolongation of the incident beam.

The deflected β-rays of radium behave in a similar manner, but the transmitted beam experiences, for the same thickness of screen, a much slighter modification. According to the experiments of M. Becquerel, the very readily deflected β-rays of radium (those with a relatively small velocity) are powerfully scattered by an aluminium screen of thickness 0·1 m.m.; but the penetrating and less deflected rays (rays of the cathode kind of great velocity) pass through this screen without being sensibly diffused, whatever be the inclination of the screen to the direction of the beam. The β-rays of great velocity penetrate without diffusion a much greater thickness of paraffin (several centimetres), and in this the curvature of the beam produced by the magnetic field can be traced. The thicker the screen, and the more absorbent the material of which it is composed, the greater is the modification of the deflected primitive beam, because, with increasing thickness of screen, diffusion occurs progressively among fresh groups of rays of increasing penetration.

The β-rays of radium experience a diffusion in passing through the air, which is very marked for readily deflected rays, but which is much slighter than that produced by equal thicknesses of solid substances. For this reason, the β-rays traverse long distances in the air.

Penetrating Power of the Radiation of Radio-active Bodies.

Since the beginning of the researches on radio-active bodies, investigations of the absorption produced by different screens upon the rays given off by these bodies have been carried on. In a previous paper on this subject I gave figures (quoted at the beginning of this work) representing the penetrating power of uranium and thorium rays. Mr. Rutherford has made a special study of the radiation of uranium, and proved it to be heterogeneous. Mr. Owens has arrived at the same results for thorium rays. When the discovery of strongly radio-active bodies immediately followed upon this, the penetrating power of their rays was also studied by various physicists (Becquerel, Meyer and von Schweidler, Curie, Rutherford). The first observations brought to light the complexity of the radiation, which seems to be a general phenomenon, and common to the radio-active bodies. In them we have sources which give rise to a variety of radiations, each of which has a power of penetration proper to itself.

Radio-active bodies emit rays which are propagated both in the air and *in vacuo*. The propagation is rectilinear; this fact is proved by the distinctness and shape of the shadows formed by interposing bodies opaque to the radiation between the source and the sensitive plate or fluorescent screen which serves as receiver, the source being of small magnitude in comparison with its distance from the receiver.

Various experiments demonstrating the rectilinear propagation of uranium, radium, and polonium rays have been made by M. Becquerel.

It is interesting to know the distance that rays can travel in air. We have found that radium emits rays which can be detected in the air at a distance of several metres from the source. In certain of our electrical determinations, the action of the source upon the air of the condenser made itself felt at a distance of between 2 and 3 metres. We have also obtained fluorescent effects and radiographic impressions at similar distances. The experiments are not easily carried out, except with very intense radio-active sources, because, independently of the absorption by the air, the action upon a given receiver varies inversely as the square of the distance from a source of small dimensions. This radiation, which travels a long distance in the case of radium, comprises rays of the cathode kind and rays which are undeflected; however, the deflected rays predominate, according to the results of the experiments already mentioned. The greater part of the radiation (α-rays) is, on the contrary, limited in air to a distance of about 7 c.m. from the source.

I made several experiments with radium enclosed in a little glass vessel. The rays emerging from the vessel, after traversing a certain space of air, were received in a condenser, which served to measure their ionising capacity by the usual electrical method. The distance, d, from the source to the condenser was varied, and the current of saturation, i, obtained in the condenser was measured. The following are the results of one of the series of determinations:—

d, c.m.	i.	$(i \times d^2 \times 10^{-1})$.
10	127	13
20	38	15
30	17·4	16
40	10·5	17
50	6·9	17
60	4·7	17
70	3·8	19
100	1·65	17

After a certain distance, the intensity of radiation varies inversely as the square of the distance from the condenser.

The radiation of polonium is only propagated in air to a distance of a few centimetres (4 to 6 c.m.) from the source of radiation.

In the case of the absorption of radiations by solid screens, we find another fundamental difference between radium and polonium. Radium emits rays capable of penetrating great thicknesses of solid matter, *e.g.*, several centimetres of lead or of glass. The rays which have passed through a great thickness of a solid body are extremely penetrating, and it is practically impossible to absorb them entirely by any material whatever. But these rays form only a small fraction of the total radiation, the greater part of which is absorbed by a slight thickness of solid matter.

Polonium emits rays which are readily absorbed, and which can only pass through extremely thin screens.

The following are figures showing the absorption produced by an aluminium lamina of thickness 0·01 m.m. This lamina was placed above and almost in contact with the substance. The direct radiation and that transmitted by the aluminium were measured by the electrical method (apparatus of Fig. 1); the current of saturation was practically obtained in every case. I have represented the activity of the radiating body by a, that of uranium being unity.

	a.	Fraction of radiation transmitted.
Chloride of barium and radium	57	0·32
Bromide ,, ,,	43	0·30
Chloride ,, ,,	1200	0·30
Sulphate ,, ,,	5000	0·29
,, ,, ,,	10,000	0·32
Metallic bismuth and polonium	—	0·22
Compounds of uranium	—	0·20
Compounds of thorium in a thin layer	—	0·38

We see that radium compounds of different nature and activity give very similar results, as I have already pointed out in the case of uranium and thorium compounds at the beginning of this work. We see also that, taking into account the whole of the radiation, and with a given absorbent screen, the different radio-active bodies can be arranged in the following decreasing order of penetrating power :—Thorium, radium, polonium, uranium.

These results are similar to those which have been published by Mr. Rutherford.

Mr. Rutherford also finds that the order is the same when air is the absorbent substance. But it is probable that this order has no absolute value, and would not be maintained independently of the nature and thickness of the screen. Experiment shows, indeed, that the law of absorption is very different for polonium and radium, and that, for the latter, the absorption of the rays of each of the three groups must be considered separately.

Polonium is particularly well adapted to the study of α-rays, because the specimens which we possess emit no other kind of rays. I made a preliminary series of experiments with extremely active recently prepared specimens of polonium. I found the absorbability of the rays to increase with increase of thickness of the matter traversed. This singular law of absorption is contrary to that known for other kinds of radiation.

I employed for this research our apparatus for the determination of electrical conductivity arranged in the following manner:—

The two plates of a condenser, P P and P' P' (Fig 8), are horizontally disposed in a metallic box, B B B B, connected to earth. The active body, A, placed in a thick metallic box, C C C C, connected with the plate P' P', acts upon the air of the condenser across a metallic sheet, T; the rays which pass through the sheet are alone utilised for producing the current, the electric field being limited by the sheet. The distance, A T, of the active body from the sheet may be varied. The field between the plates is established by means of a battery. By placing in A upon the active body different screens, and by adjusting the distance A T, the absorption of rays which travel long or short distances in the air may be determined.

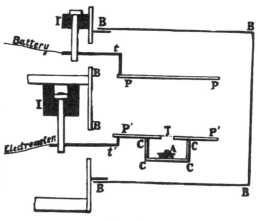

Fig. 8.

The following are the results obtained with polonium:—

For a certain value of the distance A T (4 c.m. and more), no current passes; the rays do not penetrate the condenser. When the distance A T is diminished, the appearance of the rays in the condenser is manifested somewhat suddenly, a weak current changing to one of considerable strength for a slight diminution of distance; the current then increases regularly as the active body continues to approach the sheet T.

When the active body is covered with a sheet of aluminium 1/100 m.m. thick, the absorption produced by the lamina becomes greater, the greater the distance A T.

If a second similar lamina of aluminium be placed upon the first, each absorbs a fraction of the radiation it receives, and this fraction is greater for the second lamina than for the first.

In the following table I have represented in the first line the distances in centimetres between the polonium and the sheet T; in the second line the percentage of the rays transmitted by a sheet of aluminium; in the third line the percentage of the rays transmitted by two sheets of the same aluminium:—

Distance A T	3·5	2·5	1·9	1·45	0·5
Percentage of rays transmitted by one lamina...	0	0	5	10	25
Percentage of rays transmitted by two laminæ	0	0	0	0	0·7

In these experiments the distance of the plates, P and P', was 3 c.m. We see that the interposition of the aluminium screen diminishes the intensity of the radiation to a greater degree at further distances than at nearer distances.

This effect is still more marked than the preceding figures seem to indicate. For a distance of 0·5 c.m. 25 per cent represents the mean penetration for all the rays which pass beyond this distance. If, for example, only those rays between 0·5 c.m. and 1 c.m. be comprehended, the penetration would be greater. And if the plate P be placed at a distance of 0·5 c.m. from P' the fraction of the radiation transmitted by the aluminium lamina (for A T = 0·5 c.m.) is 47 per cent, and through two laminæ it is 5 per cent of the original radiation.

I have recently performed a second series of experiments with these same specimens of polonium, the activity of which was considerably diminished, the interval of time between the two series of experiments being three years.

In the former experiments, polonium nitrite was used; in the latter, the polonium was in the state of metallic particles obtained by fusing the nitrite with potassium cyanide.

I found that the radiation of polonium had preserved its essential characteristics, and I discovered new results. The following, for different values of the distance A T, are the fractions of the radiation transmitted by a screen composed of four superposed very thin leaves of beaten aluminium.

Distance A T, in centimetres	0	1·5	2·6
Percentage of rays transmitted by the screen	76	66	39

I also found that the fraction of the radiation absorbed by a given screen increases with the thickness of the material already traversed by the radiation, but this only occurs after the distance A T has reached a certain value. When this distance is zero (the polonium being in contact with the sheet, either outside or inside the condenser), it is observed that with several similar superposed screens, each absorbs the same fraction of the radiation it receives; otherwise expressed, the intensity of the radiation diminishes therefore according to an exponential law as a function of the thickness of the material traversed, as in the case of homogeneous radiation transmitted by the lamina without changing its nature.

The following numerical results are given with reference to these experiments:—

For a distance A T equal to 1·5 c.m. a thin aluminium screen transmits the fraction 0·51 of the radiation it receives when acting alone, and the fraction 0·34 of the radiation it receives when it is preceded by another similar screen.

On the contrary, for a distance A T equal to zero, the same screen transmits in both the cases considered the same fraction of the radiation it receives, and this fraction is equal to 0·71; it is therefore greater than in the preceding case.

The following numbers indicate for a distance A T equal to 0 and for a succession of thin superposed screens, the ratio of the radiation transmitted to the radiation received for each screen:—

Series of nine very thin copper leaves.	Series of seven very thin aluminium leaves.
0·72	0·69
0·78	0·94
0·75	0·95
0·77	0·91
0·70	0·92
0·77	0·93
0·69	0·91
0·79	
0·68	

Taking into account the difficulties of the manipulation of very thin screens and of the superposition of screens in contact, the numbers of each column may be looked upon as constant; the first number only of the aluminium column indicates a greater absorption than that indicated by the following numbers.

The a-rays of radium behave similarly to the rays of polonium. These rays may be investigated almost isolated by deflecting to one side the β-rays with the magnetic field; the γ-rays seem of slight importance in comparison with the a-rays. The operation can only be carried on at some distance from the source of radiation. The following are the results of an experiment of this kind. The fraction of the radiation transmitted by a lamina of aluminium 0·01 m.m. thick is measured; this screen was placed always in the same position, above and at a little distance from the source of radiation. With the apparatus of Fig. 5, the current produced in the condenser for different values of the distance A D is observed, both with and without the screen :—

Distance A D	6·0	5·1	3·4
Percentage of rays transmitted by the aluminium	3	7	24

The rays which travel furthest in the air are those most absorbed by the aluminium. There is therefore a great similarity between the absorbable a-rays of radium and the rays of polonium.

The deflected β-rays and the undeflected penetrating γ-rays are, on the contrary, of a different nature. The experiments, notably of MM. Meyer and von Schweidler, clearly show that, considering the radiation of radium as a whole, the penetrating power of this radiation increases with the thickness of the material traversed, as is the case of Röntgen rays. In these experiments the a-rays produce scarcely any effect, being for the most part suppressed by very thin absorbent screens. Those which penetrate are, on the one hand, β-rays more or less scattered; on the other hand, γ-rays, which appear similar to Röntgen rays.

The following are the results of some of my experiments on the subject :—

The radium is enclosed in a glass vessel. The rays, which emerge from the vessel, traverse 30 c.m. of air, and are received upon a series of glass plates, each of thickness 1·3 m.m.; the first plate transmits 49 per cent of the radiation it receives, the second transmits 84 per cent of the radiation it receives, the third transmits 85 per cent of the radiation it receives.

In another series of experiments the radium was enclosed in a glass vessel placed 10 c.m. from the condenser which received the rays. A series of similar screens of lead each 0·115 m.m. thick were placed on the vessel.

The ratio of the radiation transmitted to the radiation received is given for each of the successive screens by the following numbers:—

0·40   0·60   0·72   0·79   0·89   0·92   0·94   0·94   0·97

For a series of four screens of lead, each of which was 1·5 m.m. thick, the ratio of the radiation transmitted to the radiation received was given for the successive screens by the following numbers :—

0·09   0·78   0·84   0·82

The results of these experiments show that when the thickness of the lead traversed increases from 0·1 m.m. to 6 m.m., the penetrating power of the radiation increases.

I found that, under the experimental conditions mentioned, a screen of lead 1·8 c.m. thick transmits 2 per cent of the radiation it receives; a screen of lead 5·3 c.m. thick transmits 0·4 per cent of the radiation it receives. I also found that the radiation transmitted by a thickness of lead of 1·5 m.m. consists largely of rays capable of deflection (cathode order). The latter are therefore capable of traversing not only great distances in the air, but also considerable thicknesses of very absorbent solids, such as lead.

In investigating with the apparatus of Fig. 2 the absorption exercised by an aluminium screen 0·01 m.m. thick upon the total radiation of radium, the screen being always placed at the same distance from the radiating body, and the condenser being placed at a variable distance, A D, the results obtained are the sum of those due to the three groups of the radiation. At a long distance the penetrating rays predominate, and the absorption is slight; at a short distance the α-rays predominate, and the absorption becomes less with nearer approach to the substance; for an intermediate distance the absorption passes through a maximum and the penetration through a minimum.

Distance A D	7·1	6·5	6·0	5·1	3·4
Percentage of rays transmitted by aluminium	91	82	58	41	48

Certain experiments made in connection with absorption always demonstrate a certain similarity between the α-rays and the β-rays. Thus it was that M. Becquerel discovered that the absorbent action of a solid screen upon the β-rays increases with the distance of the screen from the source, such that if the rays are subjected to a magnetic field, as in Fig. 4, a screen placed in contact with the source of radiation allows a larger portion of the magnetic spectrum to be in evidence than does the same screen placed upon the photographic plate. This variation of the absorbent effect of the screen with the distance of the screen from the source is similar to that which occurs with the α-rays; this has been verified by MM. Meyer and von Schweidler, who operated by means of the fluoroscopic method; M. Curie and I observed the same fact when working by the electrical method. However, when the radium is enclosed in a glass tube and placed at a distance from the condenser, which is itself enclosed in a thin aluminium box, it becomes a matter of indifference whether the screen be placed against the source or against the condenser; the current obtained is the same in both cases.

The investigation of the α-rays led me to the reflection that these rays behave like projectiles having a certain initial velocity, and which lose their force on encountering obstacles. These rays, moreover, travel by rectilinear propagation, as has been shown by M. Becquerel in the following experiment: — Polonium emitting rays was placed in a very narrow straight cavity hollowed in a sheet of cardboard. Thus a linear source of radiation was produced. A copper wire, 1·5 m.m. in diameter, was placed parallel and opposite to the source at a distance of 4·9 m.m. Beyond was placed a parallel photographic plate at a distance of 8·65 m.m. After an exposure of ten minutes, the geometric shadow of the wire was perfectly reproduced, with a narrow penumbra corresponding to the size of the source. The same experiment succeeded equally well when a double leaf of beaten aluminium was placed against the wire, through which the rays must pass.

There are therefore rays capable of giving perfect geometric shadows. The experiment with the aluminium shows that these rays are not scattered in traversing the screen, and that this screen does not give rise to any noticeable extent to secondary rays similar to the secondary rays of the Röntgen rays.

The experiments of Mr. Rutherford show that the projectiles which constitute the α-rays are deflected by a magnetic field, as if they were positively charged. The deflection in a magnetic field becomes less as the product

$\frac{mv}{e}$ becomes greater; m being the mass of the particle, v its velocity, and e its charge. The cathode rays of radium are but slightly deflected, because their velocity is enormous; they are, on the other hand, very penetrating, because each particle has a very small mass together with a great velocity. But particles which, with an equal charge and a less velocity, have a greater mass, would be also only slightly influenced by the action of the field, and would give rise to very absorbable rays. From the results of Mr. Rutherford's experiments, this seems to take place in the case of the α-rays.

The penetrating γ-rays appear to be of quite another nature and similar to Röntgen rays.

We have now seen how complex a phenomenon is the radiation of radio-active bodies. The difficulties of investigation are increased by the question as to whether the radiation undergoes a merely selective absorption on the part of the material, or whether a more or less radical transformation.

Little is so far known with regard to this question. If the radiation of radium be regarded as containing rays both of the cathode and Röntgen species, it might be expected to undergo transformations in traversing screens. It is known:—Firstly, that cathode rays emerging from a Crookes tube through an aluminium window are greatly scattered by the aluminium; and, further, that the passage through the screen entails a diminution of the velocity of the rays. In this way, cathode rays with a velocity equal to 1.4×10^{10} c.m. lose 10 per cent of their velocity in passing through 0·01 m.m. of aluminium. Secondly, cathode rays on striking an obstacle give rise to the production of Röntgen rays. Thirdly, Röntgen rays, on striking a solid obstacle, give rise to the production of *secondary rays*, which partly consist of cathode rays.

The existence, by analogy, of all these preceding phenomena may therefore be predicted for the rays of radio-active substances.

In investigating the transmission of polonium rays through a screen of aluminium, M. Becquerel observed neither the production of secondary rays nor any transformation into cathode rays.

I endeavoured to demonstrate a transformation of the rays of polonium by using the method of interchangeable screens. Two superposed screens, E_1 and E_2, being traversed by the rays, the order in which they are traversed should be immaterial if the passage through the screens does not transform the rays; if, on the contrary, each screen transforms the rays during transmission, the order of the screens is of moment. If, for example, the rays are transformed into more absorbable rays in passing through lead, and no such effect is produced by aluminium, then the system lead-aluminium will be more opaque than the system aluminium-lead; this takes place with Röntgen rays.

My experiments show that this phenomenon is produced with the rays of polonium. The apparatus employed was that of Fig. 8. The polonium was placed in the box, c c c c, and the absorbing screens, of necessity very thin, were placed upon the metallic sheet T.

Screens employed.	Thickness. M.m.	Current observed.
Aluminium	0·01	17·9
Brass	0·005	
Brass	0·005	6·7
Aluminium	0·01	
Aluminium	0·01	150
Tin	0·005	
Tin	0·005	125
Aluminium	0·01	
Tin	0·005	13·9
Brass	0·005	
Brass	0·005	4·4
Tin	0·005	

The results obtained prove that the radiation is modified in passing through a solid screen. This conclusion accords with the experiments in which, of two similar superposed metallic screens, the first is less absorbent than the second. From this it is probable that the transforming action of a screen increases with the distance of the screen from the source. This fact has not been verified, and the nature of the transformation has not been studied in detail.

I repeated the same experiments with a very active salt of radium; the result was negative. I only observed insignificant variations in the intensity of the radiation transmitted with interchange of the order of the screens. The following systems of screens were experimented with:—

Aluminium, thickness	... ...	0·55	m.m.
,,	,, ...	0·55	,,
,,	,, ...	0·55	,,
,,	,, ...	1·07	,,
,,	,, ...	0·55	,,
,,	,, ...	1·07	,,
,,	,, ...	0·15	,,
,,	,, ...	0·15	,,
,,	,, ...	0·15	,,
Platinum,	,, ...	0·01	,,
Lead,	,, ...	0·1	,,
Tin,	,, ...	0·005	,,
Copper,	,, ...	0·05	,,
Brass,	,, ...	0·005	,,
,,	,, ...	0·005	,,
Platinum,	,, ...	0·01	,,
Zinc,	,, ...	0·05	,,
Lead,	,, ...	0·1	,,

The system lead-aluminium was slightly more opaque than the system aluminium-lead, but the difference was not great.

Thus, I was unable to discover an appreciable transformation of the rays of radium. However, in various radiographic experiments, M. Becquerel observed very intense effects due to scattered or secondary rays, emitted by solid screens which received radium rays. Lead seemed to be the most active substance in this respect.

Ionising Action of Radium Rays on Insulating Liquids.

M. Curie has pointed out that radium rays and Röntgen rays act upon liquid dielectrics as upon air, imparting to them a certain electrical conductivity. The experiment was carried out in the following manner (Fig. 9):—

The experimental liquid is placed in a metal vessel, C D E F, into which a thin copper tube, A B, is plunged; these two pieces of metal serve as electrodes. The outer vessel is maintained at a known potential, by means of a battery of small accumulators, one pole of which is connected to earth. The tube, A B, is connected to the electrometer. When a current traverses the liquid the electrometer is kept at zero by means of a quartz electrical piezometer, which gives the strength of the current. The copper tube, M N M' N', connected to earth, serves as a guard tube, preventing the passage of the current through the air. A bulb containing the radium-barium salt may be placed at the bottom of the tube, A B; the rays act on the liquid after having penetrated the glass of the bulb and the sides of the metal tube. The radium may also be allowed to act by placing the bulb beneath the side, D E.

In working with Röntgen rays the course of the rays is through side D E.

The increase of conductivity by the action of the radium rays or the Röntgen rays seems to be produced in the case of all liquid dielectrics; but in order to determine this increase, the conductivity of the liquid itself must be so slight as not to mask the effect of the rays.

M. Curie obtained results of the same order of magnitude with both radium rays and Röntgen rays.

When investigating with the same apparatus the conductivity of air or of another gas under the action of the Becquerel rays, the intensity of the current obtained is found to be proportional to the difference of potential between the electrodes, as long as the latter does not exceed a few volts; but at higher tensions, the intensity of the current increases less and less rapidly, and the saturation current is practically attained for a tension of 100 volts.

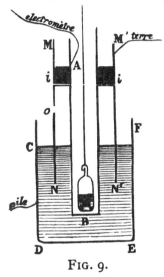

FIG. 9.

Liquids examined with the same apparatus and the same radio-active body behave differently; the intensity of the current is proportional to the tension when the latter varies between 0 and 450 volts, and when the distance between the electrodes does not exceed 6 m.m.

The figures of the following table multiplied by 10^{-11} give the conductivity in megohms per c.c. :—

Carbon bisulphide	20
Petroleum ether	15
Amylene	14
Benzine	4
Liquid air	1·3
Vaseline oil	1·0

We may, however, assume that liquids and gases behave similarly, but that, in the case of liquids, the current remains proportional to the tension up to a much higher limit than in the case of gases. It therefore seemed probable that the limit of proportionality could be lowered by using a much more feeble radiation, and this idea was verified by experiment. The radio-active body employed was 150 times less active than that which had served for the previous experiments. For tensions of 50, 100, 200, 400 volts, the intensities of the current were represented respectively by the numbers 109, 185, 255, 335. The proportionality was no longer maintained, but the current showed great variation when the difference of potential was doubled.

Some of the liquids examined are nearly perfect insulators when maintained at a constant temperature and when screened from the action of the rays. Such are liquid air, petroleum ether, vaseline oil, and amylene. It is therefore very easy to study the effect of the rays. Vaseline oil is much less sensitive to the action of the rays than is petroleum ether. This fact may have some relation to the difference in volatility which exists between these two hydrocarbons. Liquid air, which has boiled for some time in the experimental vessel, is more sensitive to the action of the rays than that newly poured in; the conductivity produced by the rays is one-fourth as great again in the former case. M. Curie has investigated the action of the rays upon amylene and upon petroleum ether at temperatures of $+10°$ and $-17°$. The conductivity due to the radiation diminishes by one-tenth of its value only, in passing from $10°$ to $-17°$.

In the experiments in which the temperature of the liquid is varied, the temperature of the radium may be either that of the surrounding atmosphere or that of the liquid; the same result is obtained in both cases. This leads to the conclusion that the radiation of radium does not vary with the temperature, and remains unaltered even at the temperature of liquid air. This fact has been verified directly by measurements.

Various Effects and Applications of the Ionising Action of the Rays Emitted by Radio-active Substances.

The rays of the new radio-active substances have a strongly ionising action upon air. By the action of radium *the condensation of supersaturated water vapour* can be easily induced, just as happens by the action of cathode rays and Röntgen rays.

Under the influence of the rays emitted by the new radio-active substances, the *distance of discharge between two metallic conductors for a given difference of potential is increased;* to put it otherwise, the passage of the spark is facilitated by these rays.

In causing conductivity, by the action of radio-active bodies, in the air in the neighbourhood of two metallic conductors, one of which is connected to earth and the other to a well-insulated electrometer, the electrometer is seen to be permanently deflected, which gives a measure of the electromotive force of the battery formed by the air and the two metals (electromotive force of contact of the two metals, when they are separated by air). This method of measurement was employed by Lord Kelvin and his students, the radiating body being uranium ; a similar method had been previously employed by M. Perrin, who was using the ionising action of Röntgen rays.

Radio-active bodies may be employed in the study of atmospheric electricity. The active substance is enclosed in a little box of thin aluminium fixed at the extremity of a metal wire connected with the electrometer. The air is made to conduct in the neighbourhood of the end of the wire, and the latter adopts the potential of the surrounding air. Radium thus replaces, with advantage, the flames or the apparatus of running water of Lord Kelvin, till now in general use for the investigation of atmospheric electricity.

Fluorescent and Luminous Effects.

The rays emitted by the new radio-active bodies cause fluorescence of certain substances. M. Curie and myself first discovered this phenomenon when causing polonium to act upon a layer of barium platinocyanide through aluminium foil. The same experiment succeeds yet more easily with barium containing radium. When the substance is strongly radio-active the fluorescence produced is very beautiful.

A large number of bodies are capable of becoming phosphorescent or fluorescent by the action of the Becquerel rays. M. Becquerel studied the effect upon the uranium salts, the diamond, &c. M. Bary has demonstrated that the salts of the metals of the alkalis and alkaline earths, which are all fluorescent under the action of luminous rays and Röntgen rays, are also fluorescent under the action of the rays of radium. Paper, cotton, glass, &c., are all caused to fluoresce in the neighbourhood of radium. Among the different kinds of glass, Thuringian glass is specially luminous. Metals do not seem to become luminous.

Barium platinocyanide is most conveniently used when

the radiation of the radio-active bodies is to be investigated by the fluoroscopic method. The effect of the radium rays may be followed at distances greater than 2 m. Phosphorescent zinc sulphide is made extremely luminous, but this body has the inconvenient property of preserving its luminosity for some time after the action of the rays has ceased.

The fluorescence produced by radium may be observed when the fluorescent screen is separated from the radium by absorbent screens. We were able to observe the illumination of a screen of barium platinocyanide across the human body. However, the action is incomparably greater when the screen is placed immediately in contact with the radium, being separated from it by no solid screen at all. All the groups of rays appear capable of producing fluorescence.

In order to observe the action of polonium, the substance must be placed close to the fluorescent screen, without the intervention of a solid screen, unless the latter be extremely thin.

The luminosity of fluorescent substances exposed to the action of radio-active bodies diminishes with time. At the same time the fluorescent substance undergoes a transformation. The following are examples:—

Radium rays transform barium platinocyanide into a brown, less luminous variety (an action similar to that produced by Röntgen rays, and described by M. Villard). Uranium sulphate and potassium sulphate are similarly altered. The changed barium platinocyanide is partially regenerated by the action of light. If the radium be placed beneath a layer of barium platinocyanide spread on paper, the platinocyanide becomes luminous; if the system be kept in the dark, the platinocyanide becomes changed, and its luminosity diminishes considerably. But if the whole be exposed to light, the platinocyanide is partially regenerated, and if the whole is replaced in darkness the luminosity reappears with vigour. By means of a fluorescent body and a radio-active body, we have therefore obtained a system which acts as a phosphorescent body capable of long duration of phosphorescence.

Glass made fluorescent by the action of radium becomes coloured brown or violet. At the same time its fluorescence diminishes. If the glass thus changed be warmed, it is decolorised, and when this occurs the glass becomes luminous. The glass has now regained its fluorescent property in the same degree as before the transformation.

Zinc sulphide, which has been exposed for a sufficient length of time to the action of radium, gradually becomes used up, and loses its phosphorescent property, whether under the action of radium or that of light.

The diamond becomes phosphorescent under the action of radium, and may thus be distinguished from paste imitations, which have only a very faint luminosity.

All the barium-radium compounds *are spontaneously luminous*. The dry anhydrous halogen salts emit a particularly intense light. This illumination cannot be seen in broad daylight, but it is easily visible in the twilight or by gas-light. The light emitted may be strong enough to read by in the dark. The light emitted emanates from the entire body of the product, whilst in the case of a common phosphorescent body, the light emanates specially from the portion of the surface illuminated. Radium products lose much of their luminosity in damp air, but they regain it on drying (Giesel). There is apparently conservation of luminosity. After many years no sensible modification is produced in the luminosity of feebly active products, kept in the dark in sealed tubes. In the case of very active and very luminous radium-barium chloride, the light changes colour after several months; it becomes more violet and loses in intensity; at the same time the product undergoes transformations; on re-dissolving the salt in water and drying it afresh, the original luminosity is restored.

Solutions of barium-radium salts, which contain a large proportion of radium, are equally luminous; this fact may be observed by placing the solution in a platinum capsule, which not being itself luminous permits of the faint luminosity of the solution being seen.

When a solution of a barium-radium salt contains crystals deposited in it, these crystals are luminous at the bottom of the solution, and much more so than the solution itself, so that they alone appear luminous.

M. Giesel has made a preparation of barium-radium platinocyanide. When this salt is newly crystallised, it has the appearance of ordinary barium platinocyanide and is very luminous. But gradually the salt becomes spontaneously coloured, taking a brown tint, the crystals at the same time becoming dichroic. In this state the salt is much less luminous, although its radio-activity is increased. The radium platinocyanide, prepared by M. Giesel, changes still more rapidly.

Radium compounds are the first example of self-luminous **bodies.**

Evolution of Heat by the Salts of Radium.

MM. Curie and Laborde have recently discovered that the salts of radium are the source of a spontaneous and continuous evolution of heat. This evolution has the effect of keeping the salts of radium at a temperature higher than that of their surroundings; an excess of temperature of 1·5° has been observed. This excess of temperature is dependent upon the thermal insulation of the body. MM. Curie and Laborde have determined the amount of heat produced in the case of radium. They found that the output is of the order of magnitude of 100 calories per grm. of radium per hour. One grm.-atom (225 grm.) of radium give rise in one hour to 22,500 cal., a quantity of heat comparable to that produced by the combustion of 1 grm.-atom (1 grm.) of hydrogen. So great an evolution of heat can be explained by no ordinary chemical reaction, more particularly as the condition of the radium remains unaffected for years. The evolution of heat might be attributed to a slow transformation of the radium atom. If this were the case, we should be led to conclude that the quantities of energy generated during the formation and transformation of the atoms are considerable, and that they exceed all that is so far known.

Chemical Effects produced by the New Radio-active Bodies.

Colourations.—The radiations of strongly radio-active bodies are capable of causing certain chemical reactions. The rays emitted by radium products exercise colouring actions upon glass and porcelain.

The colouration of glass, generally brown or violet, is very deep; it is produced in the body of the glass, and remains after removal of the radium. All glasses become coloured after a longer or a shorter interval of time, and the presence of lead is not essential. This fact may be compared to that recently observed of the colouration of the glass of vacuum tubes, after having been long in use for the production of Röntgen rays.

M. Giesel has demonstrated that the crystallised halogen salts of the alkali metals become coloured under the influence of radium, as under the action of cathode rays. M. Giesel points out that similar colourations are obtained when the salts of the alkalis are exposed to sodium vapour.

I investigated the colouration of a collection of glasses of known composition, kindly lent me for the occasion by M. Le Chatelier. I observed no great variety in the colouration. It is generally brown, violet, yellow, or grey.

It appears to be associated with the presence of the alkali metals.

With the pure crystallised alkali salts more varied and more vivid colours are obtained; the salt, originally white, becomes blue, green, yellow, brown, &c.

M. Becquerel has discovered that yellow phosphorus is transformed into the red variety by the action of radium.

Paper is changed and coloured by the action of radium. It becomes brittle, scorched, and, finally, resembles a colander perforated with holes.

Under some circumstances there is a production of ozone in the neighbourhood of very active compounds. Rays emerging from a sealed jar containing radium do not produce ozone in the air they pass through. On the contrary, a strong odour of ozone is detected when the jar is opened. In a general way, ozone is produced in the air when the latter is in direct contact with the radium. Communication by a channel, even if extremely narrow, suffices; it appears as if the production of ozone is associated with the propagation of induced radio-activity, of which we shall speak later.

Radium compounds appear to change with lapse of time, doubtless under the action of their own radiation. It was seen above that crystals of barium-radium chloride, which are colourless when formed, become gradually coloured first yellow or orange, then pink; this colouration disappears in solution. Barium-radium chloride generates oxygen compounds of chlorine; the bromide those of bromine. These slow changes generally manifest themselves some time after the preparation of the solid product, which at the same time changes in form and colour, becoming yellow or violet. The light emitted also becomes more violet.

A solution of a radium salt evolves hydrogen (Giesel).

Pure radium salts seem to undergo the same changes as those containing barium. However, crystals of the chloride, deposited in acid solution, do not become sensibly coloured after some time has elapsed, whereas crystals of barium-radium chloride, rich in radium, become deeply coloured.

Production of Thermo-luminosity.—Certain bodies, such as fluorite, become luminous when heated; they are thermo-luminescent. Their luminosity disappears after some time, but the capacity of becoming luminous afresh through heat is restored to them by the action of a spark, and also by the action of radium. Radium can thus restore to these bodies their thermo-luminescent property. Fluorite when heated undergoes a change, which is accompanied by the emission of light. If the fluorite is afterwards subjected to the action of radium, an inverse change occurs, which is also accompanied by an emission of light.

An absolutely similar phenomenon occurs when glass is exposed to radium rays. Here also a change is produced in the glass while luminous from the effect of the radium rays; this change shows itself in the colouration which appears and gradually increases. If the glass is afterwards heated, the inverse change takes place, the colour disappears, and this phenomenon is accompanied by production of light. It appears very probable that we have here a change of a chemical nature, and the production of light is associated with this change. This phenomenon may be general. It might be that the production of fluorescence by the action of radium and the luminosity of radium compounds is of necessity associated with some chemical or physical change in the substance emitting the light.

Radiographs.—The radiographic action of the new radio-active bodies is very marked. However, the method of operating should be very different with polonium and radium. Polonium acts only at very short distances, and its action is considerably weakened by solid screens; it is practically annihilated by means of a screen of slight thickness (1 m.m. of glass). Radium acts at considerably greater distances. The radiographic action of radium rays may be observed at more than 2 m. distance in air, even when the active product is enclosed in a glass vessel. The rays acting under these conditions belong to the β- and γ-groups. Owing to the differences in transparency of different materials to the rays, radiographs of different objects may be obtained, as in the case of Röntgen rays. Metals are, as a rule, opaque, with the exception of aluminium, which is very transparent. There is no noteworthy difference of transparency between flesh and bone. The operation may be carried on at a great distance and with a source of very small dimensions; and very delicate radiographs are thus produced. The beauty of the radiograph is enhanced by deflecting to one side the β-rays, by means of a magnetic field, and utilising only the γ-rays. The β-rays, in traversing the object to be radiographed, undergo a certain amount of diffusion, and thus cause a slight fog. In suppressing them, a longer time of exposure is necessary, but better results are obtained. The radiograph of an object, such as a purse, requires one day with a radiating source composed of several centigrms. of a radium salt, enclosed in a glass vessel, and placed at a distance of 1 m. from the sensitive plate, in front of which the object is placed. If the source is at a distance of 20 c.m. from the plate, the same result is obtained in one hour. In the immediate vicinity of the source of radiation, a sensitive plate is instantaneously acted upon.

Physiological Effects.

Radium rays exert an action upon the epidermis. This has been observed by M. Walkhoff and confirmed by M. Giesel, since also by MM. Becquerel and Curie.

If a celluloid or thin indiarubber capsule containing a very active salt of radium be placed upon the skin, and be left thus for some time, a redness is produced upon the skin, either immediately or at the end of some time, which is longer in proportion as the action is weaker; this red spot appears in the place which has been exposed to the action; the local change in the skin appears and acts like a burn. In certain cases a blister is formed. If the exposure was of long duration, an ulceration is produced which is long in healing. In one experiment, M. Curie caused a relatively weak radio-active product to act upon his arm for ten hours. The redness appeared immediately, and later a wound was caused which took four months to heal. The epidermis was locally destroyed, and formed again slowly and with difficulty, leaving a very marked scar. A radium burn with half-an-hour's exposure appeared after fifteen days, formed a blister and healed in fifteen days. Another burn, caused by an exposure of only eight minutes, occasioned a red spot which appeared two months after, its effect being quite insignificant.

The action of radium upon the skin can take place across metal screens, but with weakened effect.

The action of radium upon the skin has been investigated by Dr. Daulos, at the Hospital of S. Louis, as a process of treating certain affections of the skin, similar to the treatment with the Röntgen rays or the ultra-violet rays. In this respect radium gives encouraging results; the epidermis partially destroyed by the action of the radium is renewed in a healthy condition. The action of radium is more penetrating than that of light, and its use is easier than that of light or of Röntgen rays. The study of the conditions of application is of necessity rather lengthy, because the effect of the application does not at once appear.

M. Giesel has observed the action of radium upon plant

leaves. The leaves thus treated turn yellow and wither away.

M. Giesel has also discovered the action of radium rays upon the eye. If a radio-active substance be placed in the dark in the vicinity of the closed eye or of the temple, a sensation of light fills the eye. This phenomenon has been studied by MM. Himstedt and Nagel. These physicists have demonstrated that the centre of the eye is rendered fluorescent by the action of radium, and this explains the sensation of light experienced. Blind people whose retina is intact are sensitive to the action of radium, whilst those whose retina is diseased do not experience any sensation of luminosity.

Radium rays either arrest or hinder the development of colonies of microbes, but this action is not very intense.

M. Danysz has recently demonstrated the ready action of radium upon the marrow and brain. After one hour's exposure paralysis of the animals experimented upon occurred, and the latter usually died in a few days.

Influence of Temperature upon Radiation.

There is so far but little information regarding the manner of variation of the radiation of radio-active bodies with temperature. We know, however, that radiation subsists at low temperatures. M. Curie placed a glass tube containing barium-radium chloride in liquid air. The luminosity of the radio-active body persisted under these conditions. At the moment, indeed, of removing the tube from the cold bath, it appears more luminous than at the ordinary temperature. At the temperature of liquid air radium continues to cause fluorescence in the sulphates of uranium and potassium. M. Curie has verified, by electrical determinations, that the radiation, measured at a certain distance from the source, possesses the same intensity whether the radium be at the temperature of the atmosphere or of liquid air. In these experiments the radium was placed at the bottom of a tube closed at one end. The rays emerged from the tube at the open end, traversed a certain space in the air, and were received into a condenser. The action of the rays upon the air of the condenser was determined both on leaving the tube in the air and on surrounding it to a certain height with liquid air. The same result was obtained in both cases.

The radio-activity of radium persists at high temperatures. Barium-radium chloride after being fused (towards 800°) is radio-active and luminous. However, prolonged heating at a high temperature has the effect of temporarily lowering the radio-activity of the body. This decrease is very considerable; it may constitute 75 per cent of the total radiation. The decrease is less in proportion for the absorbable rays than for the penetrating rays, which are to some extent suppressed by heating. In time the radiation of the product regains the intensity and composition that it possessed before heating; this occurs after the lapse of about two months from the occasion of heating.

CHAPTER IV.

Communication of Radio-activity to Substances Initially Inactive.

During the course of our researches on radio-active substances M. Curie and I have observed that every substance which remains for some time in the vicinity of a radium salt becomes itself radio-active. In our first publication on this subject, we confined ourselves to proving that the radio-activity thus acquired by substances initially inactive is not due to the transference of radio-active particles to the surface of these substances. This is proved beyond dispute by all the experiments which will be here described, and by the laws according to which the radio-activity excited in naturally inactive bodies disappears when the latter are removed from the influence of radium.

We have given the name of *induced radio-activity* to the new phenomenon thus discovered.

In the same publication, we indicated the essential characteristics of induced radio-activity. We excited screens of different substances by placing them in the neighbourhood of solid radium salts, and we investigated the radio-activity of these screens by the electrical method. We observed the following facts:—

1. The activity of a screen exposed to the action of radium increases with the time of exposure, approaching to a definite limit according to an asymptotic law.

2. The activity of a screen which has been excited by the action of radium, and which is afterwards withdrawn from its action, disappears in a few days. This induced activity approaches zero as a function of the time, following an asymptotic law.

3. Other things being equal, the radio-activity induced by the same radium product upon different screens is independent of the nature of the screen. Glass, paper, metals, all acquire the same degree of activity.

4. The radio-activity induced in one screen by differing radium products has a limiting value which rises with the increased activity of the product.

Shortly afterwards, Mr. Rutherford published a research, which showed that compounds of thorium are capable of producing the phenomenon of induced radio-activity. Mr. Rutherford discovered for this phenomenon the same laws as those just enunciated, besides this additional important fact, that bodies charged with negative electricity become more active than others. Mr. Rutherford also observed that air passed over thorium oxide preserves a notable conductivity for about ten minutes. Air in this condition communicates induced radio-activity to inactive substances, especially to those negatively charged. Mr. Rutherford explains his experiences by the supposition that compounds of thorium, particularly the oxide, give rise to a *radio-active emanation* capable of being carried by air currents and charged with positive electricity. This emanation would be the origin of induced radio-activity. M. Dorn has repeated, with salts of barium containing radium, the experiments of Mr. Rutherford with thorium oxide.

M. Debierne has shown that actinium causes, to a marked degree, induced activity of bodies placed in its vicinity. As in the case of thorium, there is a considerable carriage of activity by air currents.

Induced radio-activity has various aspects, and irregular results are obtained when the activity of a substance in the neighbourhood of radium is excited in free air. MM. Curie and Debierne have observed, however, that the phenomenon is quite regular when operating in a closed vessel; they therefore investigated induced activity in a closed space.

Activity Induced in an Enclosed Space.

The active material is placed in a little glass jar, *a*, open at o (Fig. 11), in the centre of a closed space. Several plates, A, B, C, D, E, placed in the inclosure become radio-active after one day's exposure. The activity is the same whatever be the nature of the plate, for equal dimensions (lead, copper, aluminium, glass, ebonite, wax, cardboard, paraffin). The activity of one face of one of the plates is greater in proportion to the amount of free space about this face.

If the preceding experiment be repeated with the jar, *a*, completely closed, no activity is induced.

The radiation of radium does not directly affect the production of induced radio-activity. For this reason, in the preceding experiment the plate D, screened from the radiation by a lead plate of thickness, P P, is made as active as B and E.

Radio-activity is transmitted by the air by degrees from the radiating body to the body to be excited. It can even be transmitted to a distance by very narrow capillary tubes.

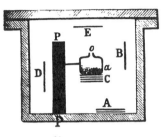

Fig. 11.

Induced radio-activity is both more intense and more regular if the solid radium salt be replaced by an aqueous solution of the same.

Liquids are capable of acquiring induced radio-activity. For example, pure water may be rendered active by placing it with a solution of a radium salt within an enclosure.

Certain substances become luminous when placed in an active enclosure (phosphorescent and fluorescent bodies, glass, paper, cotton, water, salt solutions). Phosphorescent zinc sulphide is particularly brilliant under the circumstances. The radio-activity of these luminous bodies is, however, the same as that of a piece of a metal or other body which is excited under the same conditions without becoming luminous.

Whatever be the substance made active in a closed vessel, this substance acquires an activity which increases with length of time until it attains a *limiting value*, always the same, for the same material and the same experimental arrangement.

The limit of induced radio-activity is independent of the nature and pressure of the gas inside the active enclosure (air, hydrogen, carbonic acid).

The limit of induced radio-activity for the same enclosure depends only on the quantity of radium present in the state of solution, and is apparently proportional to it.

Part played by Gases in the Phenomena of Induced Radio-activity.

Emanation.—The gases present in an enclosure containing a solid salt or a solution of a salt of radium are radio-active. This radio-activity persists when the gas is drawn off with a tube and collected in a test-tube. The sides of the test-tube become themselves radio-active, and the glass of the test-tube is luminous in the dark. The activity and luminosity of the test-tube finally completely disappear, but very gradually, and a month afterwards radio-activity may still be detected.

Since the beginning of our researches, M. Curie and I have, by heating pitchblende, extracted a strongly radio-active gas, but, as in the preceding experiment, the activity of this gas finally completely vanished.

We could discern no new ray in the spectrum of this gas; this was therefore not a case of a new radio-active gas, and we understood later that it was the phenomenon of induced radio-activity.

Thus, for thorium, radium, and actinium induced radio-activity is progressively propagated through the gases, from the radiating body to the walls of the enclosure containing it, and the exciting principle is carried away with the gas itself, when the latter is extracted from the enclosure.

When the radio-activity of radium compounds is measured by the electrical method by means of the apparatus of Fig. 1, the air between the plates is itself radio-active; however, on passing a current of air between the plates, there is no observable lowering of the intensity of the current, which proves that the radio-activity distributed in the space between the plates is of little account in comparison with that of the radium itself in the solid state.

It is quite otherwise with thorium. The irregularities which I observed in determining the radio-activity of the thorium compounds arose from the fact that at this point I was working with a condenser open to the air; the least air current caused a considerable change in the intensity of the current, because the radio-activity dispersed in the space in the vicinity of the thorium is considerable as compared with the radio-activity of the substance.

This effect is still more marked in the case of actinium. A very active compound of actinium appears much less active when a current of air is passed over the substance.

The radio-active energy is therefore contained in the gas in a special form. Mr. Rutherford suggests that radio-active bodies generate an *emanation* or gaseous material which carries the radio-activity. In the opinion of M. Curie and myself, the generation of a gas by radium is a supposition which is not so far justified. We consider the emanation as radio-active energy stored up in the gas in a form hitherto unknown.

Dissipation in Free Air of the Induced Activity of Solid Bodies.

A solid body, which has been excited by radium in an enclosed space for a sufficient length of time, and which has then been removed from the enclosure, parts with its activity in free air according to an exponential law, which is the same for all bodies represented by the following formula:—

$$I = I_0\left(ae^{-\frac{t}{\theta_1}} - (a-1)e^{-\frac{t}{\theta_2}}\right).$$

I_0 being the initial intensity of the radiation at the moment of withdrawing the plate from the enclosure; I, the intensity after time, t; a is a numerical coefficient, $a = 4·20$; θ_1 and θ_2 are time constants, $\theta_1 = 2420$ secs., $\theta_2 = 1860$ secs. After two or three hours this law becomes practically a simple exponential, and the effect of the second exponential upon the value of I is negligible. The law of dissipation is therefore such that the intensity of radiation becomes diminished to one-half its value in twenty-eight minutes. This final law may be considered as characteristic for the dissipation in an unconfined air space of the activity induced in solid bodies by radium.

Solid bodies excited by actinium lose their activity in the open air, according to an exponential law similar to the preceding, the dissipation being, however, rather slower.

Solid bodies, made active by thorium, lose their activity much more slowly; the intensity of the radiation is reduced to one-half in eleven hours.

Dissipation of Activity in a Confined Space.
Velocity of Destruction of the Emanation.

An enclosure, made active by radium and then removed from its influence, loses its activity by a law which is much less rapid than that of dissipation in the open air. The experiment may be carried out with a glass tube, rendered active internally by placing it for some time in contact with a solution of a salt of radium. The tube is then sealed in the flame, and the intensity of radiation emitted by the walls of the tube is measured while the dissipation takes place.

The law of dissipation is an exponential law. It is given very accurately by the formula.

$$I = I_0 e^{\frac{t}{\theta}}.$$

$I_0 =$ initial intensity of radiation.
$I =$ intensity of radiation at time, t.
$\theta =$ a time constant, $\theta = 4·970 \times 10^5$ secs.

The intensity of the radiation is reduced to one-half in four days.

This law of dissipation is absolutely invariable whatever be the experimental conditions (dimensions of enclosure, nature of the walls, nature of the gas within the enclosure, duration of action, &c.). The law of dissipation remains the same for any temperature between $-180°$ and $+450°$. The law is therefore altogether characteristic.

In these experiments it is the radio-active energy accumulated in the gas that maintains the activity of the walls. If the gas be withdrawn and a vacuum caused in the enclosure, we have found that dissipation of activity at once occurs in the rapid method; the intensity of radiation being reduced to one-half in twenty-eight minutes. The same result is obtained when ordinary air is substituted for the active air in the enclosure.

The law of dissipation with reduction of intensity of radiation to one-half in four days, is therefore characteristic of the disappearance of radio-active energy accumulated in the gas. By making use of the expression adopted by Mr. Rutherford, the emanation from radium may be said to disappear spontaneously as a function of the time, with reduction to one-half in four days.

The emanation from thorium is of another kind, and disappears much more rapidly. The intensity of radiation diminishes to one-half in about one minute ten seconds.

The emanation from actinium disappears still more rapidly; reduction to one-half takes place in a few seconds.

Variation of Activity of Liquids rendered Active and of Radium Solutions.

Any liquid whatever becomes radio-active when placed in an active confined space. On being removed and left freely exposed to the air, the liquid rapidly loses its activity, imparting it to the gas and solid bodies surrounding it. If a liquid thus made active be placed in a closed flask, it loses its activity much more slowly; the latter being reduced in intensity to one-half in four days, just as would a gas under similar circumstances. This fact may be explained by assuming that the radio-active energy is stored in liquids in exactly the same form as in gases (in the form of an emanation).

A solution of a radium salt behaves in a somewhat similar manner. At first, it is a remarkable fact that the solution of a radium salt placed for some time in a confined space is no more active than pure water placed in a vessel in the same enclosure, when the equilibrium of activity is established. If the radium solution be removed from the enclosure and left standing in the air in a wide-necked vessel, the activity spreads itself into space, and the solution becomes nearly inactive, though still containing radium.

If this solution be now enclosed in a stoppered flask, it gradually regains, in about a fortnight, a maximum of activity, which may be considerable. On the other hand, a liquid made active, but not containing radium, does not regain its activity in a closed flask after having been exposed to the atmosphere.

Theory of Radio-activity.

The following is, according to MM. Curie and Debierne, a very general theory which allows of the co-ordination of the results of the investigation of induced radio-activity, which results I have just stated, and which constitute facts apart from any hypothesis.

It may be said that each atom of radium acts as a constant and continuous source of energy, without actually defining the origin of this energy. The radio-active energy which accumulates in the radium tends to become dissipated in two different ways:—(1) By radiation (rays both charged and uncharged with electricity); (2) by conduction, $i.e.$, by gradual transmission to surrounding bodies in a medium of gases and liquids (production of an emanation and transformation into induced radio-activity).

The loss of radio-active energy, both by radiation and by conduction, increases with the amount of energy accumulated in the radio-active body. The system is necessarily in equilibrium when the double loss of which I have just made mention compensates the constant gain due to the action of radium. This manner of regarding the subject is similar to that in use for calorific phenomena. If in the interior of any body there is, owing to any cause, a continuous and constant evolution of heat, the heat accumulates in the body and the temperature rises until the loss of heat by radiation and conduction is in equilibrium with the constant gain of heat.

In general, except under certain special circumstances, the activity is not propagated through solid bodies. When a solution is kept in a sealed tube, the loss by radiation alone takes place, and the radiating activity of the solution is of a higher degree.

If, on the contrary, the solution stands in an open vessel, the loss of activity by conduction becomes considerable, and when the state of equilibrium is attained, the radiating activity of the solution is very feeble.

The radiating activity of a solid radium salt left exposed to the air does not sensibly diminish, because the propagation of activity by conduction not taking place through solid bodies, it is a very thin superficial layer only that produces induced radio-activity. The solution, however, of the same salt produces much more intense phenomena of induced radio-activity. With a solid salt the radio-active energy accumulates in the salt, and is dissipated chiefly by radiation. On the other hand, when the salt has been for several days in aqueous solution, the radio-active energy is divided between the salt and the water, and if separated by distillation the water carries with it a large portion of the activity, and the solid salt is much less active (ten or fifteen times) than before solution. Afterwards the solid salt gradually regains its original activity.

The preceding theory may be yet further defined by supposing the radio-activity of radium itself to be produced through the medium of the radio-active energy emitted in the form of an emanation.

Each atom of radium may be considered as a constant and continuous source of emanation. At the same moment that this form of energy is produced, it undergoes a progressive transformation into radio-active energy of the Becquerel radiation. The velocity of this transformation is proportional to the quantity of the emanation accumulated.

When a radium solution is placed within an enclosure, the emanation is able to expand into the enclosure and to spread out over the walls. Here it is, therefore, that it is transformed into a radiation, the solution giving off but few Becquerel rays; the radiation is, in some sort, *externalised*. On the other hand, with solid radium, the emanation not being able to escape readily, accumulates, and is transformed into the Becquerel radiation on the spot; this radiation therefore acquires a higher value.

If this theory of radio-activity were general, we should have to say that all radio-active bodies give rise to an emanation. Now this emission has been confirmed in the case of radium, thorium, and actinium; with the latter in particular the emission is enormous, even in the solid state. Uranium and polonium do not seem to emit any emanation, though they generate Becquerel rays. These bodies produce no induced radio-activity in an enclosed space, as do the radio-active bodies mentioned before. This fact is not in absolute contradiction to the preceding theory. If uranium and polonium were to emit emanations which become destroyed with very great rapidity, it would be very difficult to observe the carriage of such emanations by the air and the effects of induced radio-activity produced by them upon neighbouring bodies. Such a hypothesis is not improbable, since the times required for certain quantities of the emanations of radium and thorium to diminish to one-half are in the proportion of 5000 to 1. We shall see, moreover, that, under certain conditions, uranium can excite induced activity.

Another Form of Induced Radio-activity.

According to the law of dissipation in the open air of the activity induced by radium in solid bodies, the activity after one day is almost imperceptible.

Certain bodies, however, form exceptions; such are celluloid, paraffin, caoutchouc, &c. When these bodies have been acted upon to a sufficient degree, they lose their activity more slowly than the law can account for, and it is often fifteen or twenty days before the activity becomes imperceptible. These bodies appear to have the property of becoming charged with radio-active energy in the form of an emanation; they afterwards lose it gradually, causing induced radio-activity in the vicinity.

Induced Radio-activity with Slow Dissipation.

There is yet another form of induced radio-activity, which appears to be produced in all bodies which have been kept for months in an active enclosure. When these bodies are removed from the enclosure their activity at first diminishes to a very low value according to the ordinary law (diminution to one-half in half-an-hour); but when the activity has fallen to about 1/20,000 of the initial value, it diminishes no further, or at least it is dissipated very slowly, sometimes even increasing in amount. We have sheets of copper, aluminium, and glass which still retain a residual activity after six months.

These phenomena of induced radio-activity appear to be of a different kind from the ordinary ones, and show a much slower process of evolution.

A considerable time is also necessary both for the production and dissipation of this form of induced radio-activity.

Radio-activity Induced upon Substances in Solution with Radium.

When a radio-active ore containing radium is treated, with the object of extracting the radium, chemical separations are effected, after which the radio-activity is confined entirely to one of the products. In this way active products, which may be several hundred times as active as uranium, are separated from totally inactive products, such as copper, antimony, arsenic, &c. Certain other bodies (iron, lead) were never separated in an entirely inactive state. As these active bodies are concentrated, the case is no longer the same; each chemical separation no longer furnishes absolutely inactive products; all the resulting products of a separation are active in varying degrees.

After the discovery of induced radio-activity, M. Giesel was the first to attempt to excite activity in ordinary inactive bismuth by keeping it in solution with very active radium. He thus obtained radio-active bismuth, and from this he concluded that the polonium extracted from pitchblende was probably bismuth made active by the vicinity of the radium contained in the pitchblende.

I have also prepared active bismuth by keeping bismuth in solution with a very active radium salt.

The difficulties of this experiment consist in the extreme precautions which must be taken to remove all traces of radium from the solution. If we realise what an infinitesimal quantity of radium suffices to produce very considerable radio-activity in 1 grm. of material, it is difficult to believe in the possibility of sufficiently washing and purifying the active product. Each purification causes a diminution of activity of the product, whether this be due to removal of traces of radium or that the induced radio-activity is, under these circumstances, not proof against chemical reactions.

The results I obtained appear, however, to establish with certainty the fact that the activity is produced and persists after the radium is removed. On fractionating the nitrate of my active bismuth by precipitation with water from the nitric acid solution, I found that after careful purifying it fractionated like polonium, the most active portion being precipitated first.

If the purification is not complete the opposite occurs, showing that traces of radium still remain. I thus obtained active bismuth which from the manner of fractionation showed great purity and which was 2000 times as active as uranium. This bismuth diminishes in activity with lapse of time. But another portion of the same product, prepared with the same precautions, and fractionating in the same manner, preserves its activity without diminution for actually a period of about three years.

This activity is 150 times as great as that of uranium.

I have also prepared active lead and silver by leaving them in solution with radium. Generally induced radio-activity obtained in this way scarcely lessens with lapse of time, but it does not as a rule withstand many successive chemical changes of the active body.

M. Debierne made active barium by placing it in solution with actinium. This barium remains active after several chemical reactions, its activity being therefore a somewhat stable atomic property. Active barium chloride fractionates like barium-radium chloride, the more active portions being the least soluble in water and dilute hydrochloric acid. The dry chloride is spontaneously luminous: its Becquerel radiation is similar to that of barium-radium chloride. M. Debierne has prepared an active barium chloride 1000 times as' active as uranium. This barium, however, had not acquired all the characteristics of radium, for it showed none of the strongest radium lines in the spectroscope. Further, its activity diminished on standing, and after three weeks it had become one-third of its original value.

There is a wide field for research upon the radio-activity induced in substances in solution with active bodies. It appears that, according to the conditions of experiment, more or less stable forms of induced atomic radio-activity may be obtained. The radio-activity induced under these circumstances is perhaps identical with that form, which dissipates slowly, obtained by prolonged exposure at a distance in an active enclosure. We have reason to enquire to what degree induced radio-activity affects the chemical nature of the atom, and if it is able to modify the chemical properties of the latter, either temporarily or permanently.

The chemical investigation of bodies excited at a distance is rendered difficult by the fact that the induced activity is limited to a very thin superficial layer, and that, consequently, only a very small proportion of the material has been affected.

Induced radio-activity also results from leaving certain substances in solution with uranium. The experiment succeeded in the case of barium. If, as was done by M. Debierne, sulphuric acid be added to a solution containing uranium and barium, the precipitate of barium sulphate acquires radio-activity, and, at the same time, the uranium salt loses part of its activity. M. Becquerel found, after repeating this experiment several times, that almost inactive uranium was obtained. This might lead to the opinion that a radio-active body differing from uranium had been separated from the latter, its presence producing radio-activity in uranium. This, however, is not the case, for after some months the uranium regains its original activity; the precipitated barium sulphate, on the contrary, loses what it acquired.

A similar phenomenon is observed with thorium. Mr.

Rutherford precipitated a solution of a salt of thorium with ammonia; he separated off the solution and evaporated it to dryness. He thus obtained a small very active residue, and the precipitated thorium was observed to be less active than before. This active residue, to which Mr. Rutherford gives the name of *thorium* X, loses its activity after a time, whilst the thorium regains its original activity.

It appears, then, that concerning induced radio-activity all bodies do not behave in a similar manner, and that certain of them are much more readily excited than others.

Dissemination of Radio-active Particles and Induced Radio-activity of the Laboratory.

In making investigations of strongly radio-active bodies, particular precautions must be observed for obtaining delicate determinations. The different objects used in the chemical laboratory and those used for physical experiments soon acquire radio-activity, and act upon photographic plates through black paper. Dust particles, the air of the room, clothing, all become radio-active. The air of the room becomes a conductor. In our laboratory the evil has become acute, and we no longer have any apparatus properly insulated.

Special precautions must therefore be taken to avoid as much as possible the dissemination of active dust particles, and to avoid also the phenomena of induced activity.

The objects employed in chemistry should never be brought into the room where physical research is carried on, and as far as possible should be avoided any unnecessary keeping of active substances in this room. Before beginning our researches we were in the habit, in the case of electrical experiments, of making a connection between the different parts of the apparatus by insulated metallic wires, protected by metal cylinders connected to earth, which screened the wires from all outside electrical forces. In the investigation of radio-active bodies this arrangement is quite defective; the air being a conductor there is incomplete insulation between the thread and the cylinder, and the inevitable electromotive force of contact between the thread and the cylinder tends to produce a current through the air, and to cause a deflection of the electrometer. We now screen all the wires from the air by placing them inside cylinders filled with paraffin or other insulating material. It would also be advantageous in these investigations to make use of carefully enclosed electrometers.

Activity Induced Outside the Influence of Radio-active Substances.

Attempts were made to produce induced radio-activity outside the action of radio-active bodies.

M. Villard subjected to the action of the cathode rays a piece of bismuth placed as anticathode in a Crookes tube; the bismuth was thus rendered active to a very slight degree, for it required an exposure of eight days to obtain a photographic impression.

Mr. MacLennan has exposed different salts to the action of cathode rays, afterwards warming them slightly. The salts then acquired the property of neutralising bodies positively charged.

Studies of this kind are of great interest. If, by using known physical agents, it were possible to create a considerable radio-activity in bodies originally inactive, we might hope thence to discover the cause of the spontaneous radio-activity of certain substances.

Variations of Activity of Radio-active Bodies.
Effects of Solution.

The activity of polonium, as I have said above, diminishes with time. This diminution is slow, and does not take place at the same rate with different specimens. A sample of bismuth-polonium nitrate lost half its activity in eleven months, and 95 per cent in thirty-three months. Other specimens have evidenced similar diminution.

A specimen of metallic bismuth containing polonium was prepared from the nitrite, its activity after preparation being 100,000 times that of uranium. The metal is now only a body of medium radio-activity (2000 times that of uranium). Its radio-activity is determined at intervals. In six months it has lost 67 per cent of its activity.

The loss of activity does not seem to be facilitated by chemical action. In rapid chemical changes no considerable loss of activity has in general taken place.

In contrast to that which occurs with polonium, radium salts possess a permanent radio-activity which evidences no appreciable diminution after many years.

A freshly prepared radium salt in the solid state does not at first possess an activity of constant strength. Its activity increases from the time of preparation until it attains a practically constant limiting value after about one month. The opposite is the case for a solution. When freshly prepared the solution is very active, but when left exposed to the air it rapidly loses activity, and finally reaches a limiting activity which may be considerably less than the original. These variations of activity were first observed by M. Giesel. They are easily accounted for by the emanation theory. The diminution of the activity of the solution corresponds to the loss of the emanation which escapes into space; this diminution is much less when the solution is contained in a sealed tube. A solution which has lost its activity in air recovers a greater activity in a sealed tube. The time of increase of the activity of the salt which, after solution, has been recently obtained in the solid state, is that during which the emanation is being newly stored in the solid radium.

The following are some experiments on this subject :—

A solution of barium-radium chloride left exposed to the air for two days becomes 300 times less active.

A solution is enclosed in a stoppered vessel; the vessel is opened, the solution poured into a dish, and the activity determined :—

Activity immediately determined... 67
,, after two hours 20
,, ,, two days ... ... 0·25

A solution of barium-radium chloride, which has been kept open to the air, is enclosed in a sealed glass tube, and the radiation of this tube determined. The following results were observed :—

Activity determined immediately 27
,, ,, after 2 days... 61
,, ,, ,, 3 ,, ... 70
,, ,, ,, 4 ,, ... 81
,, ,, ,, 7 ,, ... 100
,, ,, ,, 11 ,, ... 100

The initial activity of a solid salt after preparation is feeble in proportion as the time of solution was long. A greater proportion of activity is then transmitted to the solvent. The following figures give the initial activity with a chloride whose limiting activity is 800, and which was kept for a given time in solution; the salt was afterwards dried, and its activity immediately determined :—

Limiting activity 800
Initial activity after solution and immediate evaporation 440
Initial activity after the salt has remained dissolved 5 days 120
Initial activity after the salt has remained dissolved 18 days 130
Initial activity after the salt has remained dissolved 32 days 114

During this experiment the dissolved salt was placed in a vessel merely covered with a watch-glass.

I made with the same salt two solutions which I kept in sealed tubes for thirteen months; one of these solutions was eight times the strength of the other:—

Initial activity of the salt in concentrated solution after evaporation 200
Initial activity of the salt in dilute solution after evaporation 100

The loss of activity of the salt is therefore greater when the amount of solvent is greater, the radio-active energy transmitted to the liquid having a greater volume of liquid to saturate and a greater space to fill. The two specimens of the same salt, which thus had a different initial activity, further increased in activity at very different rates at first; at the end of one day they had the same activity, and the increase of activity now continued in the same manner for both till the limit was reached.

When the solution is dilute the loss of activity by the salt is very rapid, as is shown by the following experiments:—Three equal portions of the same radium salt are dissolved in equal quantities of water. The first solution (*a*) is allowed to stand in contact with the air for one hour, and is then evaporated. The second solution (*b*) has a current of air passed through it for one hour, and is then evaporated. The third solution (*c*) is left exposed to the air for thirteen days, and then evaporated to dryness. The initial activity of each of the three salts is:—

For portion *a* ... ... 145·2
,, *b* ... ... 141·6
,, *c* ... ... ·102·6

The limiting activity of the same salt is about 470. We thus see that the greatest part of the effect was produced at the end of one hour. Further, the air current bubbling through solution *b* for one hour produced little effect. The proportion of salt in solution was about 0·5 per cent.

Radio-active energy in the form of an emanation is propagated with difficulty from solid radium in air; it experiences the same resistance to propagation from solid radium in a liquid. When radium sulphate is shaken with water for a whole day, its activity after the operation is practically the same as that of a portion of the same sulphate left exposed to air.

On placing the radium salt in a vacuum, all the emanation capable of displacement is withdrawn. However, the radio-activity of a radium chloride kept *in vacuo* for six days was not sensibly affected by the operation. This experiment shows that the radio-activity of the salt is principally due to the radio-active energy generated within the particles, and which is unaffected by the vacuum.

The loss of activity that radium undergoes when in solution is relatively greater for the penetrating rays than for the absorbable rays. The following are examples of this:—

A radium chloride which had reached its limit of activity, 470, was dissolved and left in solution for one hour; it was then evaporated, and its initial radio-activity determined by the electrical method. The total initial radiation was found to be equal to the fraction 0·3 of the total limiting radiation. If the determination of the intensity of radiation be made by covering the active body with an aluminium screen 0·01 m.m. thick, the initial radiation which traverses this screen is found to be only the fraction 0·17 of the limiting radiation traversing the same screen.

When the salt has been thirteen days in solution, the total initial radiation is found to be the fraction 0·22 of the total limiting radiation, and is 0·13 of the limiting radiation after traversing 0·01 m.m. of aluminium.

In the two cases the ratio of the initial radiation after solution to the limiting radiation is 1·7 times as great for the total radiation as for the radiation which has traversed 0·01 m.m. of aluminium.

It must further be mentioned that on evaporating the product after solution, it is impossible to avoid a certain period of time during which the product is in an intermediate condition, neither entirely solid nor entirely liquid. Neither can one avoid warming the product to remove the water quickly.

For these two reasons it is scarcely possible to determine the true initial activity of the product passing from solution to the solid state. In the experiments just quoted, equal quantities of the active bodies were dissolved in the same quantity of water, and the solutions were then evaporated to dryness under conditions as identical as possible, and without heating above 120° or 130°.

I investigated the law according to which the activity of a solid radium salt increases, from the moment in which the salt is obtained dry after solution to the moment in which it reaches its limit of activity. In the tables which follow, the intensity of radiation, I, is represented as a function of the time, the limiting intensity being supposed equal to 100, and the time being reckoned from the moment at which the product was dried. Table I. (Fig. 12, Curve I.) refers to the total radiation. Table II. (Fig. 12, Curve II.) refers only to the penetrating rays (rays which have traversed 3 c.m. of air and 0·01 m.m. of aluminium).

TABLE I.		TABLE I.	
Time. Days.	I.	Time. Days.	I.
0	21	0	1·3
1	25	1	19
3	44	3	43
5	60	6	60
10	78	15	70
19	93	23	86
33	100	46	94
67	100		

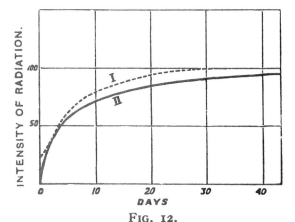

FIG. 12.

I made several other series of determinations of the same kind, but they do not absolutely agree with one another, although the general character of the curves obtained remains the same. It is difficult to obtain very regular results. It may, however, be remarked that the acquisition of activity requires more than one month for its production, and that the most penetrating rays are the most deeply affected by solution.

The initial intensity of the radiation which is able to traverse 3 c.m of air and 0·01 m.m. of aluminium is only 1 per cent of the limiting intensity, whilst the initial intensity of the total radiation is 21 per cent of the total limiting radiation.

A radium salt which has been dissolved and recently evaporated to dryness possesses the same power of causing induced activity (and, consequently, of allowing the escape of an emanation), as a specimen of the same salt which, after having been prepared in the solid state, has remained in this condition long enough to have attained its limiting radio-activity. The radiant activity of these two products is, however, quite different; the former is, for example, five times less active than the latter.

Variations of the Activity of Radium Salts on He

When a radium compound is heated, it gives off an emanation and loses activity. The more intense and the more prolonged the heating, the greater is the loss of activity. Thus, on heating a radium salt for one hour to 130°, it loses 10 per cent of its total radiation; on the other hand, heating for ten minutes to 400° produces no apparent effect. Heating to redness for several hours destroys 77 per cent of the total radiation.

The loss of activity on heating is more considerable for the penetrating than for the absorbable rays. Thus, heating for several hours destroys about 77 per cent of the total radiation, but the same amount of heating destroys nearly the whole (99 per cent) of the radiation that traverses 3 c.m. of air and 0·01 m.m. of aluminium. If barium-radium chloride be kept fused for several hours (towards 800°) 98 per cent of the radiation capable of traversing 0·3 m.m. of aluminium is destroyed. The penetrating rays may be considered as no longer in existence after intense and prolonged heating.

When a radium salt has lost part of its activity by heating, the diminution is not lasting; the activity of the salt is spontaneously regenerated at the ordinary temperature, and approaches a certain limiting value. I have observed the curious fact that this limit is higher than the limiting activity of the salt before being heated—this, at least, is the case with the chloride. I give examples of this:—A specimen of barium-radium chloride which, after having been prepared in the solid state, has long since attained its limiting activity, possesses a total radiation represented by the number 470, and a radiation capable of traversing 0·01 m.m. of aluminium, represented by the number 157. This specimen is heated to redness for several hours. Two months after the heating it attains its limit of activity with a total radiation equal to 690, and a radiation through 0·01 m.m. of aluminium, equal to 227. The total radiation and the radiation transmittted by aluminium are therefore increased respectively in the ratios $\frac{690}{470}$ and $\frac{227}{156}$. These two ratios are practically equal to one another, and are equivalent to 1·45.

A specimen of radium-barium chloride which, after having been prepared in the solid state, has attained a limiting activity of 62, is maintained in a state of fusion for some hours; the fused product is then powdered. The product regains a new limiting activity equal to 140, which is twice as great as that to which it was able to attain when prepared in the solid state without having been sensibly heated during evaporation.

I have investigated the law of increase of activity of radium compounds after heating. The following are the results of two series of determinations:—The figures of Table I. and II. represent the intensity of the radiation (I) as a function of time, the limiting intensity being supposed equal to 100, and the time being reckoned from the close of the heating. Table I. (Fig. 13, Curve I.) refers to the total radiation of a specimen of barium-radium chloride. Table II. (Fig. 13, Curve II.) relates to the penetrating radiation of a specimen of barium-radium sulphate, the intensity of the radiation which traversed 3 c.m. of air and 0·01 m.m. of aluminium having been determined. The two products were subjected to a bright red heat for seven hours.

Table I.		Table II.	
Time. Days.	I.	Time. Days.	I.
0	16·2	0	0·8
0·6	25·4	0·7	13
1	27·4	1	18
2	38	1·9	26·4
3	46·3	6	46·2
4	54	10	55·5
6	67·5	14	64
10	84	18	71·8
24	95	27	81
57	100	36	91
		50	95·5
		57	99
		84	100

I made several other series of determinations, but the results did not agree well.

The effect of heating does not persist when the heated radium compound is dissolved. Of two specimens of the same radium compound of activity 1800, one was strongly heated and its activity thereby reduced to 670. The two portions being now dissolved and left in solution for twenty hours, their initial activities in the solid state were 460 for the not heated portion and 420 for the heated one; the difference between the two portions was therefore not considerable. But if the two products do not remain for a sufficient length of time in solution—if, for example, they are evaporated to dryness, immediately after solution—the not heated product is much more active than the heated one; a certain time is necessary in the dissolved state for the effects of heating to disappear. A product of activity 3200 was heated, and its activity thereby reduced to 1030. This product and a similar portion which had not been heated were simultaneously dissolved, and the two then immediately evaporated to dryness. The initial activity was 1450 for the not heated portion, and 760 for the heated one.

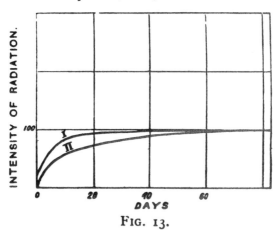

FIG. 13.

In the case of solid radium salts the capacity for exciting induced radio-activity is largely affected by heating. During heating radium compounds give off a larger amount of emanation than at the ordinary temperature; but on being cooled to the ordinary temperature, not only is their radio-activity much less than before heating, but their capacity for inducing activity is much diminished. During the time that follows the heating, the radio-activity of the product increases, and may even exceed the original value. The induction capacity is also partially re-established; however, after prolonged heating to redness, this capacity is almost entirely destroyed without spontaneous re-appearance afterwards. The induction capacity may be restored to the radium salt by dissolving it in water, and drying it in the oven at a temperature of 120°. This seems to have the effect of leaving the salt in a peculiar physical condition, in which the emanation is given off with much less facility than is the case with the same solid product not heated to a high temperature, and it follows naturally that the salt attains a higher limit of activity than that which it possessed before heating. To transform the salts into the physical condition proper to it before heating, it suffices to dissolve it and to evaporate it to dryness without heating it above 150°.

The following are numerical examples of the above :—

I represent by a the limit of induced activity produced in a closed vessel upon a plate of copper by a specimen of barium-radium carbonate of activity 1600.

Suppose $a = 100$ for the not heated product. We find—

1 day after heating	$a = 3\cdot3$
4 days ,, ,,	$a = 7\cdot1$
10 ,, ,, ,,	$a = 15$
20 ,, ,, ,,	$a = 15$
37 ,, ,, ,,	$a = 15$

The radio-activity of the product had diminished 90 per cent by heating, but one month afterwards the original value was regained.

The following is an experiment of the same kind made with a barium-radium chloride of activity 3000. The induction capacity is determined in the same manner as before.

For the product not heated $a = 100$.

Induction capacity of the product after being heated to redness for three hours :—

2 days after heating	$2\cdot3$
5 ,, ,,	$7\cdot0$
11 ,, ,,	$8\cdot2$
18 ,, ,,	$8\cdot2$
Induction capacity of the unheated substance which has been dissolved and then dried at 150°	92
Induction capacity of the heated substance which has been dissolved and then dried at 150°	105

Theory of Interpretation of the Causes of Variations of Activity of Radium Salts after Solution and after Heating.

The facts previously indicated may be in part explained by the theory according to which the energy of radium is produced in the form of an emanation, which is then transformed into the energy of radiation. When a radium salt is dissolved, the emanation produced by it spreads beyond the solution, and causes radio-activity outside the source from which it proceeds; when the solution is evaporated, the solid salt obtained is but slightly active, because it contains only a small amount of emanation. Gradually the emanation is accumulated in the salt, the activity of which rises to a limiting value, which is reached when the production of the emanation by the radium compensates the loss by external emission and by local transformation into Becquerel rays.

When a radium salt is heated, the external emission of the emanation is greatly increased, and the phenomena of induced radio-activity are more intense than when the salt is at the ordinary temperature. But when the salt returns to the ordinary temperature it is exhausted, as is the case after being dissolved, and contains but a small amount of emanation, its activity having become greatly reduced. Gradually the emanation accumulates afresh in the solid salt, and the radiation increases.

It may be said that radium gives rise to a constant generation of any emanation—part of which escapes to the exterior, the remainder being transformed in the radium itself into Becquerel rays. When radium is raised to a red heat, it loses the greater part of its capacity to cause the induction of activity; otherwise stated, the evolution of the emanation is lessened. Consequently, the proportion of the emanation utilised in the radium itself should be greater, and the substance attains a higher limit of radio-activity.

We will endeavour to establish theoretically the law of rise of activity of a solid radium salt which has been dissolved or has been heated. We will assume that the intensity of radiation of radium is, at each instant, proportional to the quantity of emanation, q, present in the radium. We know that the emanation is spontaneously destroyed according to a law such that, at each instant—

$$q = q_0 e^{-\frac{t}{\theta}} \quad \ldots \quad \text{I,}$$

q_0 being the amount of the emanation at the moment of starting the observation, and θ the time constant, equal to $4 \cdot 97 \times 10^6$ secs.

Now let Δ be the evolution of the emanation by radium, a quantity which I will assume constant. Let us consider what would occur if no emanation were escaping to the exterior. The emanation generated would then be completely utilised by the radium for the production of the radiation. We have from Formula 1—

$$\frac{dq}{dt} = -\frac{q_0}{\theta} e^{-\frac{t}{\theta}} = -\frac{q}{\theta} ;$$

and consequently, in the state of equilibrium, the radium would contain a certain quantity of emanation, Q, such that—

$$\Delta = \frac{Q}{\theta} \qquad \ldots \ldots \ldots 2,$$

and the radiation of the radium would then be proportional to Q.

Let us suppose the radium placed in the circumstances under which it gives off the emanation to the exterior; this is obtained by dissolving the radium compound or by heating it. The equilibrium will be disturbed, and the activity of the radium diminished. But as soon as the cause of the loss of emanation has been abolished (the body being restored to the solid state or the heating having ceased), the emanation is accumulated afresh in the radium and we have a period during which the evolution, Δ, surpasses the velocity of destruction, $\frac{q}{\theta}$. We then have—

$$\frac{dq}{dt} = \Delta - \frac{q}{\theta} = \frac{Q-q}{\theta},$$

from which—

$$\frac{d}{dt}(Q-q) = -\frac{Q-q}{\theta},$$

$$Q-q = (Q-q_0)e^{-\frac{t}{\theta}} \qquad \ldots \ldots 3,$$

q_0 being the amount of emanation present in the radium at time $t = 0$.

According to Formula 3, the excess of the quantity of emanation, Q, contained by the radium in a state of equilibrium above the quantity, q, contained at a given moment, decreases as a function of the time according to an exponential law, which is also the law of the spontaneous disappearance of the emanation. The radiation of radium being proportional to the amount of emanation, the excess of the intensity of the limiting radiation above the actual intensity should decrease as a function of the time by the same law; the excess should thus diminish to one-half in about four days.

The preceding theory is incomplete, since the loss of emanation to the exterior has been neglected. It is also difficult to determine the manner in which this acts as a function of the time. In comparing the results of experiment with those of this incomplete theory, there is found to be no satisfactory agreement; the conviction is, however, retained that the theory in question is partially true. The law by which the excess of limiting activity above the actual activity diminishes to one-half in four days represents approximately the course of the renewal of activity after heating for ten days. In the case of the renewal of activity after solution, the same law appears to hold approximately for a certain period of time, which begins two or three days after evaporation to dryness and continues for ten or fifteen days. The phenomena are otherwise complex; the theory sketched out does not explain the reason of the suppression of the penetrating rays in greater proportion than the absorbable rays.

Nature and Cause of the Phenomena of Radioactivity.

From the beginning of research upon the radio-active bodies, and when the properties of these bodies were yet hardly known, the spontaneity of their radiation presented itself as a problem having the greatest interest for physicists. To-day we have advanced considerably in the understanding of radio-active bodies, and are able to isolate one of very great power, viz., radium. With the object of making use of the remarkable properties of radium, a profound investigation of the rays emitted by radio-active bodies is indispensable; the various groups of rays under investigation present points of similarity with the groups of rays existing in Crookes tubes: cathode rays, Röngten rays, canal rays. The same groups of rays are found in the secondary radiation produced by Röntgen rays, and in the radiation of bodies which have acquired radio-activity by induction.

But if the nature of the radiation is actually better known, the cause of this spontaneous radiation remains a mystery, and the phenomena always presents itself to us as a profound and wonderful enigma.

The spontaneously radio-active bodies, and in the first place radium, are sources of energy. The evolution of energy, to which they give rise, is manifested by Becquerel radiation, by chemical and luminous effects, and by the continuous generation of heat.

The question often arises as to whether energy is created within the radio-active bodies themselves, or whether it is borrowed by them from external sources. No one of the numerous hypotheses arising from these two points of view has yet received experimental confirmation.

The radio-active energy may be assumed to have been initially accumulated and then gradually dissipated, as happens in the case of long continued phosphorescence. We imagine the evolution of radio-active energy to correspond to a transformation of the nature of the atom of the active body; the fact of the continuous generation of heat by radium speaks in favour of this hypothesis. The transformation may be assumed to be accompanied by a loss of weight and by an emission of material particles constituting the radiation. The source of energy may yet be sought in the energy of gravitation. Finally, we may imagine that space is constantly traversed by radiations yet unknown, which are arrested in their course by radio-active bodies and transformed into radio-active energy.

Many reasons are adduced for and against these different views, and most often attempts at experimental verifications of the conclusions drawn from these hypotheses have given negative results. The radio-active energy of uranium and radium apparently neither becomes exhausted nor varies appreciably with lapse of time. Demarçay examined spectroscopically a specimen of pure radium chloride after a five months' interval, and observed no change in the spectrum. The principal barium line, which was visible in the spectrum indicating the presence of a trace of barium, had not increased in intensity during the interval, showing therefore that there was no transformation of radium into barium to an appreciable extent.

The variations of weight announced by M. Heydweiller in radium compounds cannot yet be looked upon as established facts.

Elster and Geitel found that the radio-activity of uranium is not affected at the bottom of a mine-shaft 850 m. deep; a layer of earth of this thickness would therefore not affect the hypothetical primary radiation which would be excited by the radio-activity of uranium.

We have determined the radio-activity of uranium at midday and at midnight, thinking that if the hypothetical primary radiation had its origin in the sun it would be partly absorbed in traversing the earth. The experiment showed no difference in the two determinations.

Conclusions.

I will define, in conclusion, the part I have personally taken in the researches upon radio-active bodies.

I have investigated the radio-activity of uranium compounds. I have examined other bodies for the existence of radio-activity, and found the property to be possessed by thorium compounds. I have made clear the atomic character of the radio-activity of the compounds of uranium and thorium.

I have conducted a research upon radio-active substances other than uranium and thorium. To this end I investigated a large number of substances by an accurate electrometric method, and I discovered that certain minerals possess activity which is not to be accounted for by their content of uranium and thorium.

From this I concluded that these minerals must contain a radio-active body different from uranium and thorium, and more strongly radio-active than the latter metals.

In conjunction with M. Curie, and subsequently with MM. Curie and Bémont, I was able to extract from pitchblende two strongly radio-active bodies—polonium and radium.

I have been continuously engaged upon the chemical examination and preparation of these substances. I effected the fractionations necessary to the concentration of radium, and I succeeded in isolating pure radium chloride. Concurrently with this work, I made several atomic weight determinations with a very small quantity of material, and was finally able to determine the atomic weight of radium with a very fair degree of accuracy. The work has proved *that radium is a new chemical element.* Thus the new method of investigating new chemical elements, established by M. Curie and myself, based upon radio-activity, is fully justified.

I have investigated the law of absorption of polonium rays, and of the absorbable rays of radium, and have demonstrated that this law of absorption is peculiar and different from the known laws of other radiations.

I have investigated the variation of activity of radium salts, the effect of solution and of heating, and the renewal of activity with time, after solution or after heating.

In conjunction with M. Curie, I have examined different effects produced by the new radio-active substances (electric, photographic, fluorescent, luminous colourations, &c.).

In conjunction with M. Curie, I have established the fact that radium gives rise to rays charged with negative electricity.

Our researches upon the new radio-active bodies have given rise to a scientific movement, and have been the starting-point of numerous researches in connection with new radio-active substances, and with the investigation of the radiation of the known radio-active bodies.